MAKING THE TRANSITION TO
UNIVERSITY
CHEMISTRY

MAKING THE TRANSITION TO
UNIVERSITY
CHEMISTRY

MICHAEL CLUGSTON

formerly Tonbridge School

MALCOLM STEWART

University of Oxford

FABRICE BIREMBAUT

Caen, France

OXFORD

UNIVERSITY PRESS

Great Clarendon Street, Oxford, OX2 6DP,
United Kingdom

Oxford University Press is a department of the University of Oxford.
It furthers the University's objective of excellence in research, scholarship,
and education by publishing worldwide. Oxford is a registered trade mark of
Oxford University Press in the UK and in certain other countries

Published in the United States of America by Oxford University Press
198 Madison Avenue, New York, NY 10016, United States of America

British Library Cataloguing in Publication Data
Data available

Library of Congress Control Number: 2021934816

ISBN 978–0–19–875715–3

Printed in Great Britain by
Bell & Bain Ltd., Glasgow

LEARNING WITH THIS BOOK

We have written *Making the Transition to University Chemistry* to support you as you make the transition from pre-university to university-level study. We have presented each topic in a consistent way, using a progressive approach: we remind you of familiar ideas; we then encourage you to take the next step in your learning; and finally we invite you to take a deeper look at certain topics.

Familiar ideas

You already know a number of very important concepts about chemistry from your pre-university studies. For example, you know that Gibbs energy explains which reactions can happen, and you understand a number of important organic mechanisms. We present these familiar ideas in the form of **expanded bullet points**. Because these ideas should be familiar, we provide only the most important conclusions. If you find that you do not remember one of these ideas, look back at the pre-university textbook you used to get a more detailed explanation. **You should expect your university lecturers to assume that you understand all the expanded bullet points.**

For some of these concepts, the approach taken pre-university differs from that at university: we then add **marginal notes** (some of which may be in the main text area) to alert you to the issue in question. For example, it is frequently important to specify rather more carefully at university when a particular concept can be rigorously applied.

- **Simple molecular solids** are held together by intermolecular forces:
  - Intermolecular forces are relatively weak, so the melting and boiling points of simple molecular solids are generally low.
  - Very small molecules such as H_2, N_2, and O_2 are gases at room temperature.
  - Molecules with stronger intermolecular forces may be liquids or even solids.
    - For the halogens, dispersion forces increase as the number of electrons increases: at room temperature, F_2 and Cl_2 are gases, Br_2 is a liquid, and I_2 is a solid, as shown in Figure 4.8.
  - For the alkanes the dispersion forces increase as the number of carbon atoms increases, so the first four alkanes are gases, then come liquids, and eventually solids. Figure 4.9 shows the increasing boiling points of the alkanes.
  - The trend of increasing boiling points as the size of the molecule increases is disrupted by the existence of hydrogen bonding for ammonia, water, and hydrogen fluoride (which makes their boiling points

Taking the Next Step

Taking the Next Step sections build on the familiar ideas by providing significant expansion by way of more detailed explanations, which will allow you to make links to work that will be important at university. Some of you might have come across some of these ideas before in your particular pre-university course, but many of you will probably not have seen the ideas before.

We have chosen to present those ideas which we believe will be most useful during your early university studies. **We would strongly encourage you to study each of these sections**, as they will help you significantly when you first arrive at university.

TAKING THE NEXT STEP 8.1

Unusual oxidation numbers for oxygen and hydrogen

While oxygen, when combined in a compound, almost always has oxidation number −2, there are a few important exceptions. In the molecule **hydrogen peroxide** H_2O_2, the two oxygen atoms are bonded together and hence the electrons in that bond are shared equally. As each of the two hydrogens has Ox(H) = +1, the two oxygens must add up to −2 to make the total zero for this neutral molecule; hence each oxygen has Ox(O) = −1. (The same applies to compounds such as sodium peroxide Na_2O_2.) **Potassium superoxide** KO_2 on the other hand has Ox(K) = +1 and, unusually, Ox(O) = −1/2.

In the compound **oxygen difluoride** OF_2, oxygen combines with the only element with a higher electronegativity and so has a *positive* oxidation number of +2 with Ox(F) = −1, as always for fluorine when it is in a compound.

Hydrogen forms compounds called **hydrides** with alkali metals such as sodium (sodium hydride is NaH). In hydrides, hydrogen is the *more* electronegative atom and hence has an oxidation number Ox(H) = −1.

A Deeper Look

A Deeper Look sections typically deal with more difficult concepts because the mathematical level is higher or the explanation is more complicated. For example, in Chapter 9 we explain how to calculate the actual magnitude of the effect of temperature on the equilibrium constant, rather than simply being able to say whether it goes up or down.

Some of the concepts we present in these sections are among the most exciting in the subject, and those who relish extension work will particularly enjoy these sections. However, **it is not essential to understand all of them**, as we believe few if any university courses will expect you to know these concepts before you arrive.

We wish you every success in your future studies of chemistry!

> **A DEEPER LOOK 9.1**
>
> The Third Law of thermodynamics
>
> We can trace the reason why standard entropies are absolute values back to the **Third Law of thermodynamics**, which states that
>
> **The entropy of a perfectly crystalline substance at absolute zero is zero.**
>
> A perfect crystal has all the constituent particles in their correct places, so there is no disorder at all and thus its entropy is zero. The fact that measurements start from zero means that there is no delta in the symbol for the standard entropy ($S^{\ominus}$), unlike the case for the standard enthalpy change of formation ($\Delta_f H^{\ominus}$).

Acknowledgements

MJC is very grateful to Prof Neil Allan for commenting on many chapters throughout the book, and to Prof Nick Norman for commenting on the whole of the inorganic section. Their insightful advice was invaluable. MJC would like to thank Prof Peter Atkins for providing a number of computer images. He is also grateful to his son, John Clugston, for drawing almost all the organic mechanisms along with some of the organic structures and for his help with editing. MJC would also like to thank Jon Crowe for his guidance, encouragement, wise advice, and skilful editing throughout the development of the book. We also thank Mark Walker, who produced many of the figures used throughout the book.

ATOMIC STRUCTURE

Atoms bond together to form compounds, so before investigating compounds we need to study their building-blocks, namely atoms. Figure 1.1 shows a scanning tunnelling microscope (STM) image of palladium atoms on a graphite surface.

1.1 THE NUCLEAR MODEL

- An **atom** consists of electrons surrounding a very small, central, positively charged **nucleus**, which contains protons and neutrons.
- The **electron** has a charge of *exactly* the same magnitude, but opposite sign (negative), to that of the **proton** (positive). The **neutron** is neutral.
- The mass of the electron is much smaller than the masses of the proton or the neutron. The latter two have almost, *but not quite*, the same mass, as shown in Table 1.1.

FIGURE 1.1 This STM image shows palladium atoms on a graphite surface.

TABLE 1.1 The charges and approximate relative masses of electrons, protons, and neutrons.

	Electron	Proton	Neutron
Relative charge	−1	+1	0
Relative mass	1/1840	1	1*

*The mass of the neutron is 1.00138 times the mass of the proton.

A **nuclide** is an isotope with a specified mass number. We measure atomic masses relative to the nuclide ^{12}C; radiocarbon dating uses the nuclide ^{14}C.

- The number of protons in the nucleus of an atom equals the **atomic number Z** of the element, which can be shown by a left subscript (almost always omitted) in its full symbol.
  - So for chlorine ($_{17}Cl$), the number of protons equals 17. *All* atoms of chlorine contain 17 protons.
- The number of neutrons can, however, vary. **Isotopes** of an element have the same number of protons but different numbers of neutrons.
- The number of neutrons equals the **mass number** of the isotope minus the atomic number. The mass number is shown by a left superscript in its full symbol.
  - The isotope ^{35}Cl (chlorine-35) contains 35 − 17 = 18 neutrons.
  - The isotope ^{37}Cl (chlorine-37) contains 37 − 17 = 20 neutrons.
  - Two of the isotopes of uranium ($_{92}U$), ^{235}U and ^{238}U, both contain 92 protons, but the former has 235 − 92 = 143 neutrons whereas the latter has 238 − 92 = 146 neutrons.
- In the neutral atom, the number of electrons equals the number of protons.
- An element's chemistry is dominated by the number and arrangement of its electrons.

1.2 THE MASSES OF ATOMS

- We can measure the masses of atoms with a **mass spectrometer**. A **mass spectrum**, such as the one shown in Figure 1.2, is a plot of the **relative abundance** of each isotope against its **mass-to-charge ratio m/z**.
- The **relative atomic mass A_r** of an element is the average mass of one atom of the element relative to 1/12th the mass of one atom of carbon-12. A_r has no units. The **relative isotopic mass** is the value for a specific isotope. Because of the slight difference in mass between a proton and a neutron (Table 1.1), relative isotopic masses are *not* integers (unlike mass numbers).
- We can calculate an element's relative atomic mass as a **weighted mean** of its relative isotopic masses.

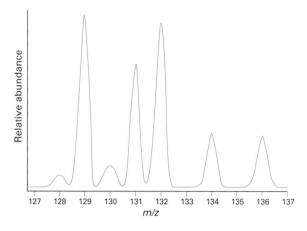

FIGURE 1.2 The mass spectrum for naturally occurring xenon. The horizontal axis is the mass-to-charge ratio *m/z*. As all ions are assumed to be unipositive, this measures the isotopic mass. The vertical axis shows the relative abundance.

- Lead has four main isotopes with mass numbers 204 (1.4 per cent), 206 (24.1 per cent), 207 (22.1 per cent), and 208 (52.4 per cent). So we can find the relative atomic mass as follows:

$$A_r = \left(\frac{1.4}{100} \times 204\right) + \left(\frac{24.1}{100} \times 206\right) + \left(\frac{22.1}{100} \times 207\right) + \left(\frac{52.4}{100} \times 208\right)$$
$$= 207.2$$

Taking the Next Step 1.1 explains why this typical pre-university calculation is usually *slightly* wrong.

TAKING THE NEXT STEP 1.1

Why does the A_r calculation usually fail to give the exact answer?

The pre-university use of *mass numbers* in the calculation of relative atomic mass is a simplification, because each isotope's relative isotopic mass is not an integer. To get an exact answer, we must use relative isotopic masses in place of mass numbers.

For chlorine, 75.8 per cent is ^{35}Cl and 24.2 per cent is ^{37}Cl so using their mass numbers gives

$$A_r = \left(\frac{75.8}{100} \times 35\right) + \left(\frac{24.2}{100} \times 37\right) = 35.48 \,(2\,\text{dp})$$

Using their relative isotopic masses (34.969 and 36.966) gives

$$A_r = \left(\frac{75.8}{100} \times 34.969\right) + \left(\frac{24.2}{100} \times 36.966\right) = 35.45 \,(2\,\text{dp})$$

The latter figure agrees with the exact relative atomic mass of chlorine.

1.3 ELECTRONIC STRUCTURE

- **Electronic structure** is frequently called **electronic configuration**.

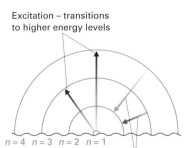

Excitation – transitions to higher energy levels

Transitions to lower energy levels emit energy in the form of electromagnetic radiation

$n = 4$ $n = 3$ $n = 2$ $n = 1$

FIGURE 1.3 The Bohr model of an atom. The red and blue transitions explain the red and blue lines in the Balmer series, see Figure 1.4.

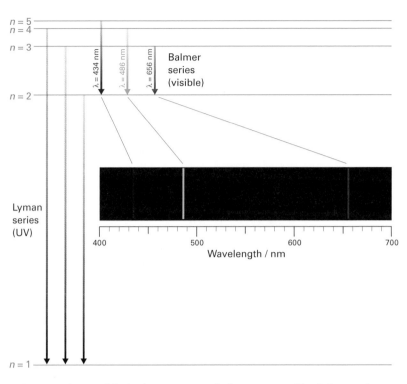

FIGURE 1.4 Part of the hydrogen atom emission spectrum. The Balmer series results from transitions from higher energy levels down to energy level $n = 2$.

$$1 \text{ nanometre } (1 \text{ nm}) = 1 \times 10^{-9} \text{ m}.$$

- The **Bohr model** (1913) proposed that the electron in a hydrogen atom could only orbit the nucleus in one of several **shells** at fixed distances from the nucleus, rather like planets orbiting round the sun: shell $n = 1$ is closest to the nucleus, as seen in Figure 1.3. This explained in detail the hydrogen atom emission spectrum shown in Figure 1.4.

- However, all attempts to extend the Bohr model to other atoms failed. We now know that shells must be subdivided into **subshells**, so that there are more energy levels available to explain the larger number of emission lines for atoms such as sodium. (Shell $n = 3$, for example, has three subshells.) Figure 1.5 shows that sodium has three lines around 600 nm, where hydrogen has *no* lines.

- We can explain the various lines around 600 nm by transitions between the different subshells. Figure 1.6 (a **Grotrian diagram**) shows how the three lines in Figure 1.5 arise.

FIGURE 1.5 The part of the sodium atom emission spectrum around 600 nm.

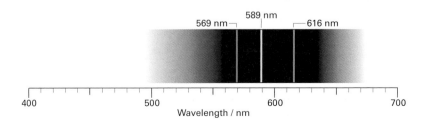

1.4 THE SCHRÖDINGER EQUATION

- We can explain how subshells arise using the **Schrödinger equation**, introduced by Erwin Schrödinger in 1926.

- A consequence of the introduction of his theory was that the *certainty* Bohr proposed for the distances at which electrons orbited around the nucleus had to be replaced with less exact knowledge. So we choose to use the word 'orbital' to replace Bohr's 'orbit'. We now focus on the **electron density**, which describes the probability of finding an electron in a particular volume. Figure 1.7 describes the electron density in the 1s orbital: the legend contains important information.

- A Deeper Look 1.3 at the end of this chapter explains the history of quantum mechanics, which led to the Schrödinger equation.

- The mathematical solutions of the Schrödinger equation show that in shell $n = 3$ there are three subshells: the 3s subshell (containing just one orbital), the 3p subshell (containing three orbitals), and the 3d subshell (containing five orbitals).

- The shapes of the three subshells are significantly different: s orbitals are spherical (circular in 2D, Figure 1.7e), p orbitals have two lobes (Figure 1.8), and four of the five d orbitals have four lobes (Figure 1.9): see Figure 2.16 for the shape of the fifth.

- The orbitals normally encountered are 1s, 2s, 2p, 3s, 3p, (3d/4s), and 4p in order of increasing energy: the 3d and 4s orbitals have similar energies. **This order needs to be learnt**. A Deeper Look 1.1 explains the reason for the notation s, p, d, and f.

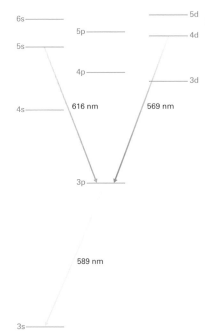

FIGURE 1.6 A Grotrian diagram (drawn accurately to scale) showing some of the electronic transitions responsible for the emission spectrum of sodium. Notice how the energies of the subshells get closer to each other as the energy increases.

The solutions of the Schrödinger equation show that in shell $n = 4$ there are four subshells: the 4s subshell (one orbital), the 4p subshell (three orbitals), the 4d subshell (five orbitals), and the 4f subshell (seven orbitals). See A Deeper Look 1.2.

You will learn much more about the Schrödinger equation and how to solve it during your university course.

A DEEPER LOOK 1.1

Why are the orbitals called s, p, d, and f?

This naming has a historical basis in the study of the sodium atom emission spectrum shown in Figures 1.5 and 1.6. The **sharp** series (including the 616 nm line) arises from transitions from an **s** orbital to a p orbital. The **principal** series (including the 589 nm line) arises from transitions from a **p** orbital to an s orbital: it was so named because the same transitions can be seen in absorption, as the outermost electron in sodium is in the 3s orbital. The **diffuse** series (including the 569 nm line) arises from transitions from a **d** orbital to a p orbital; the **fundamental** series (not shown) arises from transitions from an **f** orbital to a d orbital.

A DEEPER LOOK 1.2

Quantum numbers

We can only solve the Schrödinger equation for the hydrogen atom for certain values of the energy and can label the full set of solutions with just three quantum numbers. In general, a **quantum number** is an integer (in some cases, a half-integer, A Deeper Look 26.1) that labels the state of a

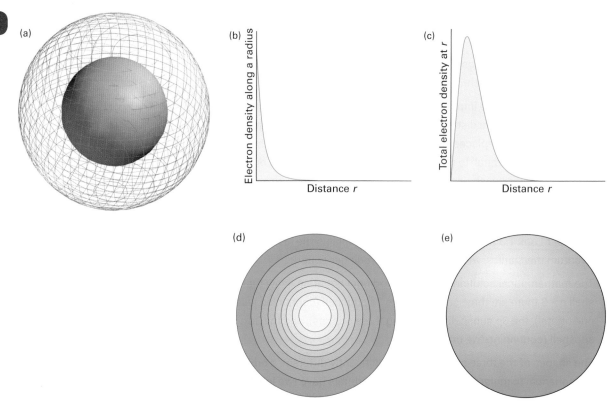

FIGURE 1.7 The electron density in the 1s orbital is spherical. (a) The solid figure joins points of equal electron density; the wire-frame joins points where the electron density is ten times lower. (b) As you move out from the nucleus along any radius of the sphere that makes up the 1s orbital, the electron density falls steadily. (c) The total electron density *at a particular distance* from the centre of the sphere (rigorously called the **radial distribution function**) varies as shown here, because all points on the surface of the sphere (of area $4\pi r^2$) are at the same distance from the nucleus. (We call the distance at which the maximum electron density occurs the **Bohr radius**, the distance Bohr calculated for the first shell.) (d) A cross-section through the electron density shows concentric circles, each successive circle including 10 per cent more of the total electron density. The outer circle includes 90 per cent. (e) To sketch an s orbital in the rest of the book, we will draw the 90 per cent contour.

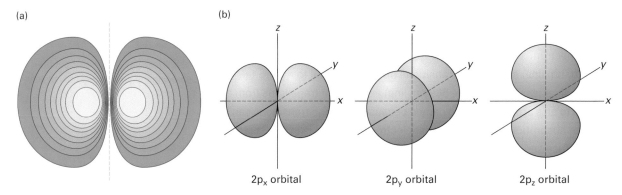

FIGURE 1.8 (a) A cross-section through the electron density of a 2p orbital: each successive contour includes 10 per cent more of the total electron density. Note that there is zero electron density along the vertical line in (a), which we call a **nodal plane**. (b) The 90 per cent boundary surfaces (i.e. the 'shapes') of the $2p_x$, the $2p_y$, and the $2p_z$ atomic orbitals.

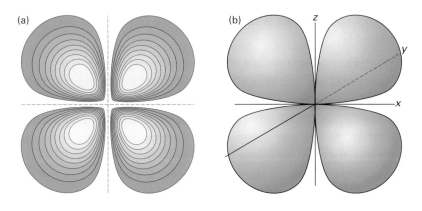

(a)　(b)

FIGURE 1.9 There are five 3d orbitals. This diagram shows (a) a cross-section through the electron density: each successive contour includes 10 per cent more of the total electron density; (b) the shape of the $3d_{xz}$ orbital.

system. The first is the **principal quantum number n**, which matches the label used by Bohr for his orbits:

- n increases in integer steps: 1, 2, 3, …

The second quantum number that arises from the solutions is the **orbital angular momentum quantum number l**. The *magnitude* of the orbital angular momentum (in units of the Planck constant, A Deeper Look 1.3, divided by 2π) is given by the formula $\sqrt{l(l+1)}$. An s orbital has $l = 0$, a p orbital has $l = 1$, and a d orbital has $l = 2$.

- For any particular value of n, l can take all integer values from 0 to $n - 1$.

Therefore, the third shell $n = 3$ has three subshells: 3s ($l = 0$), 3p ($l = 1$), and 3d ($l = 2$). The fourth shell $n = 4$ has four subshells: 4s ($l = 0$), 4p ($l = 1$), 4d ($l = 2$), and 4f ($l = 3$).

The Schrödinger equation provides a natural explanation for the existence of subshells.

The third quantum number that arises from the solutions is the **magnetic quantum number m_l**. The *projection* of the orbital angular momentum onto a magnetic field (in units of the Planck constant divided by 2π) is given by m_l.

- For any particular value of l, m_l can take all integer values from $-l$ to l.

  - There is **only one s orbital** for any value of n, as m_l must equal 0 when $l = 0$.

  - There are **three p orbitals** when $n \geq 2$, as m_l can equal -1, 0, or 1 when $l = 1$.

  - There are **five d orbitals** when $n \geq 3$, as m_l can equal -2, -1, 0, 1, or 2 when $l = 2$.

  - There are **seven f orbitals** when $n \geq 4$, as m_l can equal -3, -2, -1, 0, 1, 2, or 3 when $l = 3$.

The Schrödinger equation provides a natural explanation for the number of orbitals in each subshell and therefore the fact that each row of the f block (Figure 11.1) contains 14 elements.

This is also called the *Aufbau* principle (*Aufbau* is German for 'construction').

- The **building-up principle** states that electrons fill the lowest-energy orbitals first.

- The **Pauli exclusion principle** states that each orbital can hold a maximum of two electrons: when two electrons fill an orbital, their spins must be paired. In an orbital an electron is either spin-up or spin-down.

  > At university you will learn that **spin** is the intrinsic angular momentum a particle possesses.

- For a neutral atom, the atomic number Z gives the number of electrons; calcium ($Z = 20$) has the electronic structure $1s^2\ 2s^2\ 2p^6\ 3s^2\ 3p^6\ 4s^2$. Krypton ($Z = 36$) has the electronic structure $1s^2\ 2s^2\ 2p^6\ 3s^2\ 3p^6\ 3d^{10}\ 4s^2\ 4p^6$.

We call this rule **Hund's rule**; we call orbitals that have the same energy **degenerate**.

- When several orbitals in a subshell (such as p or d) have the same energy, they fill first with one electron in each, with *parallel* spins.

- We use the same convention for the electronic structure of *ions*.

  - For example, the outermost two electrons of calcium can be lost to form the calcium ion Ca^{2+} which has the electronic structure $1s^2\ 2s^2\ 2p^6\ 3s^2\ 3p^6$, the same as the electronic structure for argon.

  - The chloride ion Cl^- has *gained* one electron in addition to the 17 already present, so its electronic structure is the same as that of Ca^{2+} (and Ar).

  - For **transition metal ions** the situation is complicated by the fact that the 3d and 4s electrons have similar energy in the atom. The rule (for which unfortunately there is no simple explanation) is that the *4s electrons are lost first*, leaving 3d as the subshell in which we place any other electrons.

  - Iron ($Z = 26$) forms $Fe^{2+}\ 1s^2\ 2s^2\ 2p^6\ 3s^2\ 3p^6\ 3d^6$ and $Fe^{3+}\ 1s^2\ 2s^2\ 2p^6\ 3s^2\ 3p^6\ 3d^5$.

1.5 IONIZATION ENERGY

- The **first ionization energy** is the minimum energy required to remove one electron from an isolated atom in the gas phase:

$$E(g) \rightarrow E^+(g) + e^-(g)$$

  - See Sections 11.3, 11.4, and 12.2 for the trends in physical properties such as ionization energies, atomic radii, etc., across periods and down groups in the periodic table.

  - Taking the Next Step 1.2 explains why there are several numbering systems for the periodic table.

- **Successive ionization energies** provide strong evidence for the existence of shells. Figure 1.10 shows that large breaks occur in the value of the ionization energy for sodium after one and nine electrons are removed. This pattern shows that the outermost shell contains one electron, the middle shell contains eight electrons, and two electrons occupy the innermost shell.

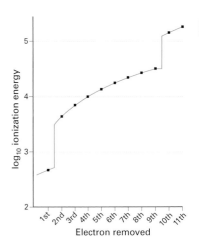

FIGURE 1.10 A plot of successive ionization energies for the sodium atom, $_{11}$Na $1s^22s^22p^63s^1$.

TAKING THE NEXT STEP 1.2

Why are there several numbering systems for the groups?

Numbering systems for the groups in the periodic table have evolved over time. The current system used at university is based on the Schrödinger equation (Section 1.4). Because there are five d orbitals, three p orbitals, and one s orbital, each of which can hold two electrons, we expect eighteen elements in the later rows of the table. Hence we number the groups from Group 1 (the alkali metals) to Group 18 (the noble gases): these are *Arabic* numerals.

In the early twentieth century, *Roman* numerals were used for the eight most important groups: Group I for the alkali metals to Group VIII for the noble gases. However, early chemistry courses in schools often translate the Roman numerals into something more familiar and therefore write these as 1 to 8 rather than I to VIII. Thus most pre-university specifications describe the halogens as Group 7 rather than Group VII or Group 17, which can lead to obvious confusion and ambiguity.

Because there are seven orbitals in an f subshell (Section 1.4), the additional two rows usually shown at the bottom of the table (Figure 11.1) contain 14 elements each.

A DEEPER LOOK 1.3

A brief history of quantum mechanics

The idea that energy can only be supplied in finite chunks called **quanta** (singular: quantum) was used for the first time in 1900 when Max Planck explained why the quantity of black-body radiation varied as the frequency emitted changed. In doing so he introduced a constant, now named for him, that relates the energy E of a **photon** (the particle found in electromagnetic radiation) to its frequency f:

$$E = hf$$

where h is the **Planck constant**.

The physics community was really impressed when Albert Einstein (in 1905) was able to use the same equation to explain a completely different phenomenon (the photoelectric effect, the emission of an electron from a metal when electromagnetic radiation hits it). *No* emission occurs below a certain threshold frequency. The status of the Planck constant became even more prominent when Niels Bohr (in 1913) modelled the hydrogen atom as one electron orbiting one proton, held in orbit by electrostatic attraction. Bohr was able to explain the hydrogen atom emission spectrum quantitatively. These three breakthrough calculations transformed the subject significantly and were rewarded with the 1918 (Planck), 1921 (Einstein), and 1922 (Bohr) Nobel Prizes in Physics.

Even more dramatic changes occurred very soon after that. In his PhD thesis (in 1924), Prince Louis de Broglie made the astonishing claim that **fundamental particles such as electrons behaved like waves**. Almost all

The symbol for frequency used at university by chemists is usually the Greek letter nu, ν. (Physicists tend to stick to f.)

FIGURE 1.11 (a) The diffraction pattern produced by a beam of X-rays passing through aluminium foil. The rows of atoms in the metal act like the slits in a diffraction grating. (b) The diffraction pattern resulting from a beam of electrons.

(a)
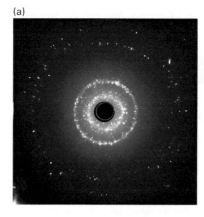

(b)
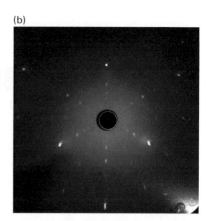

FIGURE 1.12 The winning team – Ernest Rutherford and research colleagues at the Cavendish Laboratory, Cambridge, in 1920. James Chadwick (discoverer of the neutron) is at the extreme left of the middle row. J. J. Thomson (electron) and Rutherford (proton) are three and four places, respectively, to Chadwick's left. G. P. Thomson is sitting next to Chadwick.

G. P. Thomson was the son of J. J. Thomson, who had won the Nobel Prize for proving the electron was a particle. Now his son shared the Prize for proving it was a wave! We describe this as **wave-particle duality**. Figure 1.12 illustrates both G. P. Thomson and J. J. Thomson, along with Ernest Rutherford and James Chadwick (who discovered the proton and neutron respectively).

scientists were highly sceptical, until just three years later electrons were diffracted (a characteristic behaviour of waves). The experiments were performed in the UK by G. P. Thomson and in the US by Davisson and Germer. Figure 1.11 shows diffraction patterns produced by X-rays and by electrons.

A small number of theoreticians believed de Broglie's idea *before* it was proved correct by experiment. Prompted by Victor Henri, Erwin Schrödinger produced (in 1926) his **wave equation**, which describes how the electron would behave *if it were a wave*. Not only could the Schrödinger equation re-produce the same values for the energy Bohr had found for hydrogen, but it provided a natural explanation for the existence of subshells and hence for the detailed shape of the periodic table, which the Bohr model could not do.

We now have images (such as Figure 1.13) that make it easier to believe that electrons behave like waves.

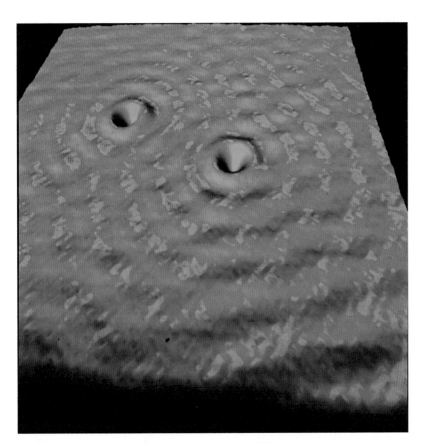

FIGURE 1.13 An STM image showing interference between electrons, confirming their wave nature. The two point defects (probably impurity atoms created in the preparation of the sample) scatter the surface state electrons, resulting in circular standing wave patterns.

2

BONDING AND MOLECULAR SHAPE

2.1 COVALENT BONDING

- A **covalent bond** occurs when atoms **share a pair of electrons**. The nuclei of the bonded atoms attract the shared electron pair. Figures 2.1 and 2.2 show **Lewis structures**, which use dots to represent electrons (**dot-and-cross diagrams** are similar but show the source of the electrons being shared; however, once shared the electrons are indistinguishable).
  - A **double** covalent bond, as shown in Figure 2.2(a), consists of two shared electron pairs.
  - A **triple** covalent bond, as shown in Figure 2.2(b), consists of three shared electron pairs.
- Covalent bonding between *identical* atoms shares the electron pair *equally*, giving a symmetrical electron density evenly spread between the two atoms, as shown in Figure 2.3.

Modern theories of covalent bonding are dominated by molecular orbital theory. Taking the Next Step 2.1 explains how molecular orbital theory describes the molecules H_2 and O_2.

Shared pair
of electrons

FIGURE 2.1 The Lewis structure for the chlorine molecule.

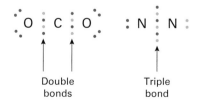

Double
bonds

Triple
bond

FIGURE 2.2 The Lewis structures for (a) carbon dioxide, and (b) nitrogen.

(a)

(b)

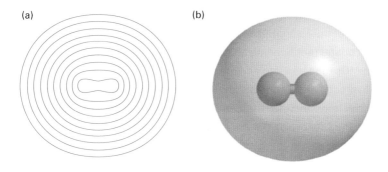

FIGURE 2.3 Two ways of picturing the electron density in the hydrogen molecule H_2: (a) A contour map – showing lines of equal electron density; (b) An electron density diagram – a computer-generated plot using theoretically calculated values of electron density.

2

TAKING THE NEXT STEP 2.1

Molecular orbital theory for H_2 and O_2

A diatomic molecule forms when the atomic orbitals of two atoms **overlap**. Once we accept the idea that electrons in atoms behave like waves (A Deeper Look 1.3), we need to consider what happens when the orbitals of two atoms overlap as an interaction between waves, as shown in Figure 2.4. When two atomic orbitals (AOs) overlap, electron density redistributes, creating two molecular orbitals. **Molecular orbitals (MOs)** are the solutions of the Schrödinger equation for molecules, in the same way that atomic orbitals are the solutions for atoms.

When two waves overlap, there can be constructive interference (when the waves are *in* phase) and destructive interference (when the waves are *out of* phase). **Constructive interference** (the same colour shows that the two atomic orbitals are in phase) will *increase* the electron density between the nuclei, which lowers the energy. Figure 2.5 shows that constructive interference of the 1s orbitals of two hydrogen atoms forms a **bonding molecular orbital (bonding MO)**. We call this orbital a **sigma (σ) orbital** because when looking around the internuclear axis the electron density is completely symmetrical, as in an s orbital: σ is the Greek letter corresponding to s.

Destructive interference (the different colour shows that the two atomic orbitals are out of phase) will *decrease* the electron density between the

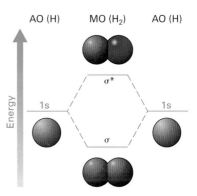

FIGURE 2.5 In-phase overlap (same colour) gives a bonding MO (labelled σ); out-of-phase overlap gives an antibonding MO (labelled σ*). When widely separated, we do not know their relative phases, as indicated by the purple colour.

Furthest apart

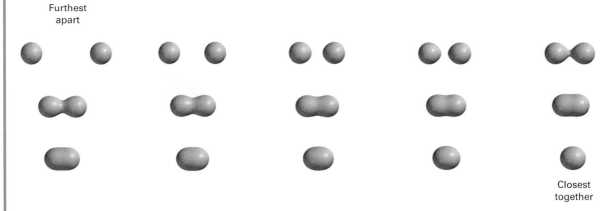

Closest together

FIGURE 2.4 Electron density diagrams showing two hydrogen atoms approaching each other and forming a hydrogen molecule. Note the shift in the distribution of electron density as the nuclei get closer.

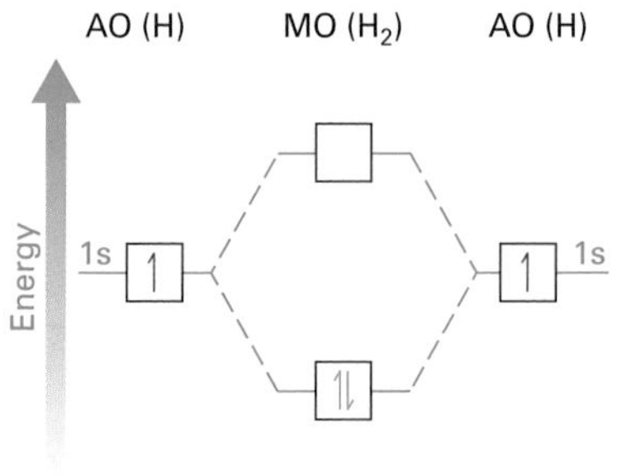

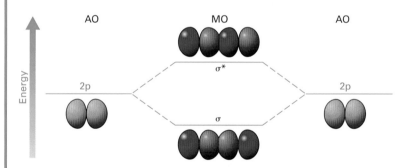

FIGURE 2.6 An energy level diagram for H_2 showing two electrons in the bonding MO of a hydrogen molecule. Note that the antibonding MO is empty.

nuclei, which increases the energy. Figure 2.5 also shows that destructive interference of the 1s orbitals of two hydrogen atoms forms an **antibonding molecular orbital (antibonding MO)**, called σ* (**sigma-star**). So the two AOs form two MOs (one bonding and one antibonding).

Each hydrogen atom can contribute one electron. Figure 2.6 shows that the two electrons then fill the bonding molecular orbital. The two spins must be paired following the Pauli exclusion principle (Section 1.4). The shared electron pair in the bonding MO constitutes the covalent bond between the two hydrogen atoms.

The *hypothetical* molecule He_2 does not exist because both the bonding and antibonding molecular orbitals are filled as there are now four electrons to be added. (When rigorous mathematical calculations are done, it turns out that the antibonding MO rises in energy more than the bonding MO falls in energy.) There must be more electrons in bonding orbitals compared with antibonding orbitals for a molecule to be formed.

The big difference between oxygen and hydrogen is that for oxygen, in addition to the overlap of s orbitals, p orbitals can now overlap. (The 1s orbitals overlap very little because they are very close to their parent atoms.) The 2s orbitals overlap to produce σ and σ* MOs just as for hydrogen.

However, p orbitals can overlap in two different ways. Figure 2.7 shows that they can overlap **end-on**, which creates σ and σ* MOs (once again the overlapping lobes are in phase for the σ and out of phase for the σ*).

However, Figure 2.8 shows that they can also overlap **sideways**, which creates **pi (π)** and **pi-star (π*)** MOs (once again the overlapping lobes are

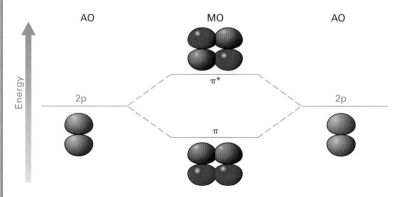

FIGURE 2.7 End-on in-phase overlap of p orbitals forms a sigma (σ) bonding MO.

FIGURE 2.8 Sideways in-phase overlap of p orbitals forms a pi (π) bonding MO.

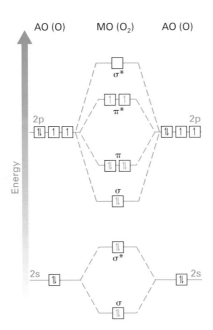

AO (O)   MO (O₂)   AO (O)

FIGURE 2.9 Oxygen's MOs: the two unpaired electrons make oxygen paramagnetic.

in phase for the π and out of phase for the π*). A π MO has a **nodal plane** (a plane in which the electron density is zero) along the internuclear axis: π is the Greek letter corresponding to p.

Considering all the MOs that can be formed, we can construct an energy level diagram for O_2, as shown in Figure 2.9. We place the electrons in the orbitals following the building-up principle and Hund's rule (Section 1.4). This means that there will be two *unpaired* electrons in the π* MOs. Substances that have unpaired electrons are **paramagnetic**, which means that they are attracted into a magnetic field.

Figure 2.10 illustrates the experimental proof that O_2 is indeed paramagnetic. This established the molecular orbital theory as a superior

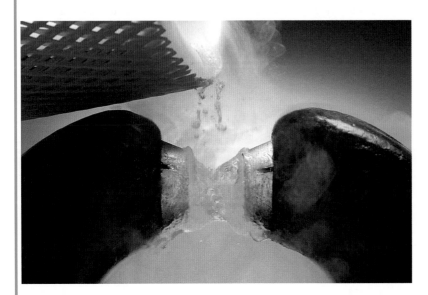

FIGURE 2.10 Liquid oxygen poured between the poles of a strong magnet sticks to them.

2

description of bonding to wthe simple Lewis structure, which suggested that all the electrons in O_2 were paired. The number of bonds predicted is, however, the same: a double bond. This is because

> **bond order** = (the number of electrons in bonding orbitals minus the number of electrons in antibonding orbitals) divided by two.

> The superoxide ion O_2^- is found in solids such as potassium superoxide (A Deeper Look 11.1) KO_2. O_2^- has *one more* electron in one of the π* MOs, making it paramagnetic but not as strongly so as in O_2 itself, as there is one less unpaired electron. The species O_2^+ (the dioxygenyl ion) was historically important for the discovery of the first noble-gas compound. O_2^+ has *one less* electron in one of the π* MOs than in O_2, making it paramagnetic to the same extent as O_2^-. Because the electron removed comes from an *antibonding* orbital, the bond in O_2^+ is shorter and stronger than that in O_2 (the bond in O_2^- is longer and weaker still).

2.2 POLAR COVALENT BONDING

- **Electronegativity** is a rough measure of the power of an atom to attract a shared electron pair in a covalent bond. (There are several scales: that introduced by Linus Pauling is the most common.) Figure 2.11 shows that electronegativity generally increases across a period and decreases down a group. (Noble gases have no electronegativity value.)
  - The two elements with the highest electronegativities are fluorine and oxygen.
- Covalent bonding between *different* atoms shares the electron pair *unequally*, giving an unsymmetrical electron density skewed towards the atom with the higher electronegativity. Such a bond is called a **polar covalent bond**. Figure 2.12 shows the electron density in the compound gallium arsenide.
  - The **more electronegative** atom gains a **partial negative charge** (indicated by a **δ−** sign); the **less electronegative** atom gains a **partial positive charge** (indicated by a **δ+** sign).

FIGURE 2.11 The elements and their electronegativity values arranged in a periodic table.

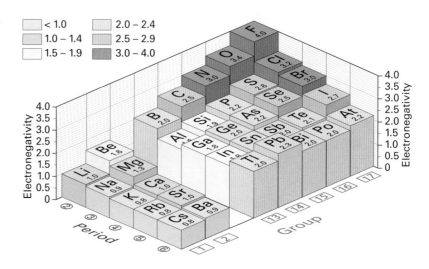

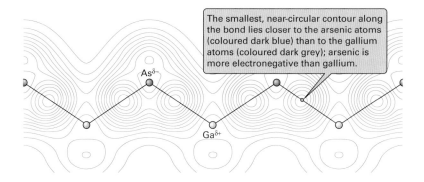

> The smallest, near-circular contour along the bond lies closer to the arsenic atoms (coloured dark blue) than to the gallium atoms (coloured dark grey); arsenic is more electronegative than gallium.

$As^{\delta-}$

$Ga^{\delta+}$

FIGURE 2.12 The experimental electron density for gallium arsenide.

2

- The larger the electronegativity difference between the bonded atoms, the larger the partial charges are and the more polar the bond is.

Taking the Next Step 2.2 describes how we can extend molecular orbital theory to cover polar covalent bonding.

TAKING THE NEXT STEP 2.2

Molecular orbital theory of polar covalent bonding

When two atoms are not identical, the atomic orbitals that overlap are at different energy levels. The two orbitals *will* still overlap and form bonding and antibonding MOs; see Taking the Next Step 2.1. However, the bonding MO will resemble the *lower*-energy atomic orbital more than the higher-energy one, so the electron density will not be symmetrical but skewed towards the atom with the higher electronegativity. Figure 2.13 shows that, for the molecule HF, fluorine is the more electronegative atom and hence has the lower orbital energy. Fluorine has a partial negative charge, while hydrogen has a partial positive charge:

$$H^{\delta+} - F^{\delta-}$$

An **electrostatic potential map**, as shown in Figure 2.13, superimposes the partial charges on top of the electron density. A blue colour indicates a partial positive charge, and a red colour indicates a partial negative charge.

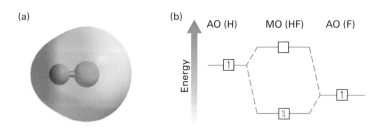

(a)

(b)

AO (H)   MO (HF)   AO (F)

Energy

FIGURE 2.13 (a) The electron density for HF is significantly higher near fluorine. The electrostatic potential map shows that there is a partial negative charge on fluorine, indicated by the red colour. The blue colour shows the partial positive charge on hydrogen. (b) The bonding MO is closer in energy to fluorine's atomic orbital (AO).

2.3 DELOCALIZATION

- **Delocalization** occurs when *more than two atoms* are involved in the bonding.

- The most familiar molecule that has delocalized bonding is **benzene**, C_6H_6.

  - **Benzene is unusually stable** because of its delocalized bonding (Section 19.1).

  - Figure 2.14 shows that the six p orbitals of the six carbon atoms overlap to form the **lowest-energy delocalized bonding molecular orbital**. (Section 19.1 shows the other two bonding molecular orbitals in benzene.)

- Other examples of delocalized bonding are as follows:

  - The ions from carboxylic acids are delocalized (Section 23.1). So, for example, ethanoic acid CH_3COOH forms the **ethanoate ion** $CH_3CO_2^-$. Figure 2.15 shows that the two oxygen atoms are identical in $CH_3CO_2^-$.

(a)   (b)

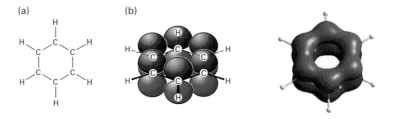

FIGURE 2.14 (a) Each carbon atom uses its 2s and two of its 2p atomic orbitals to form the *σ framework* of the benzene molecule. (b) The remaining 2p atomic orbitals all combine together to form a set of delocalized π molecular orbitals. The six remaining valence electrons fill three bonding molecular orbitals, each of which is delocalized (together constituting a *delocalized π cloud*). The lowest-energy orbital arises from overlap of all six orbitals in phase.

FIGURE 2.15 The two oxygen atoms in (a) ethanoic acid are distinguishable, but those in (b) ethanoate ion are identical, as indicated by the equally intense red colour.

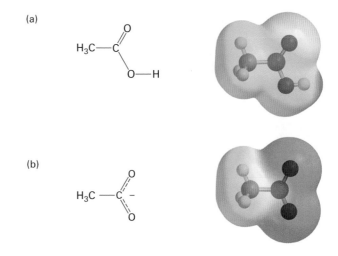

2

FIGURE 2.16 The two delocalized π orbitals in chromate(VI) ion. The unusually shaped d orbital (the one with a torus) is called d_{z^2}. The other d orbital has the usual four lobes (Section 1.4).

- The π bonds in carbonate, nitrate, and chromate(VI) ions, for example, are delocalized. Figure 2.16 shows the shapes of the delocalized orbitals in **chromate(VI) ion**.
- **Graphite** has delocalized bonding within each layer (Section 4.4), which explains its 5,000 times greater electrical conductivity along the layer compared with perpendicular.

Note that all the examples of delocalization so far discussed have involved **delocalized pi (π) bonds**. A Deeper Look 2.1 introduces the idea of **delocalized sigma (σ) bonds**.

A DEEPER LOOK 2.1

Can sigma bonds be delocalized?

For many years, the idea of delocalized *sigma* bonds was not considered. Historically the first definite example was found in the structure of the molecule diborane B_2H_6. While a second-year Oxford undergraduate, Christopher Longuet-Higgins proposed a structure involving two **three-centre, two-electron bonds**, as shown in Figure 2.17; the Nobel Laureate Linus Pauling supported a different structure. When the definitive experiments (electron diffraction and X-ray crystallography) were done, the undergraduate was proved correct!

In the 1980s, spectroscopic measurements made by astronomers identified an unexpected species, the ion H_3^+, in the ionosphere of Jupiter and later in the interstellar medium. Subsequent computer calculations identified the triangular shape as optimal, as shown in Figure 2.18. It is very easy to count the number of atoms (3) and the number of electrons (3 − 1, because of the single positive charge, making 2). This is incontrovertibly a **three-centre, two-electron delocalized sigma bond**, as hydrogen has only the 1s orbital. It is the simplest possible example of such a bond.

Once we acknowledge the concept of delocalized sigma bonds, we can find other examples. Section 17.2 explains that the sigma bonds in methane are delocalized.

It is well known that metals have a delocalized 'sea' of electrons (Section 4.4). Sodium (in the s block) has only one s electron in its valence shell and so can only form sigma bonds. So sodium must also have delocalized sigma bonds.

(a)

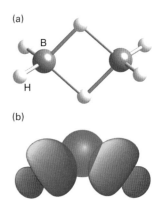

(b)

FIGURE 2.17 (a) The structure of diborane B_2H_6. (b) One of the two three-centre, two-electron delocalized sigma bonds.

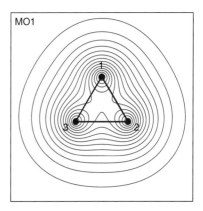

MO1

FIGURE 2.18 The structure of H_3^+.

2.4 VALENCE-SHELL ELECTRON-PAIR REPULSION

- We can work out the shapes of simple molecules using the **valence-shell electron-pair repulsion (VSEPR) theory** introduced by Ronald Nyholm and Ronald Gillespie. The total number of **bond pairs** (shared electron pairs, double bonds being treated in the same way as single bonds) and **lone pairs** (electron pairs confined to one atom) in the **valence** (outermost) **shell** determines the basic shape of the molecule:
  - **Two** electron pairs adopt a **linear** shape.
  - **Three** electron pairs adopt a **trigonal planar** shape.
  - **Four** electron pairs adopt a **tetrahedral** shape.
  - **Five** electron pairs adopt a **trigonal bipyramidal** shape.
  - **Six** electron pairs adopt an **octahedral** shape.

  Figure 2.19 shows these shapes.

FIGURE 2.19 VSEPR explains the shapes adopted by electron pairs: (a) linear; (b) trigonal planar; (c) tetrahedral; (d) trigonal bipyramidal; (e) octahedral.

(a)   (b)   (c)   (d)   (e)

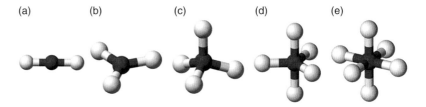

- If **all electron pairs are bond pairs, they repel each other equally**; the observed shape is the same as the basic shape, and the bond angles are as expected from geometry, as shown in Figure 2.20:
  - **linear** 180°
  - **trigonal planar** 120°
  - **tetrahedral** 109° 28' (the **tetrahedral angle**, about 109.5°)
  - **trigonal bipyramidal** 90° and 120°
  - **octahedral** 90°
- If **lone pairs are present**, the bond angles are altered because the repulsion between two lone pairs is greater than the repulsion between a lone pair and a bond pair, which is in turn greater than the repulsion between two bond pairs (**LP/LP > LP/BP > BP/BP**). The observed shape is also different from the basic shape, as the observed shape focuses only on the distribution of the atoms.
  - Figure 2.21 shows that one lone pair and three bond pairs, as is the case for ammonia (NH_3), form a **pyramidal** molecule.
  - Figure 2.22 shows that two lone pairs and two bond pairs, as is the case for water (H_2O), form a **V-shaped** or **angular** molecule.
    - For both ammonia and water, the exact angle between the bonds can rigorously only be predicted by VSEPR to be *smaller* than

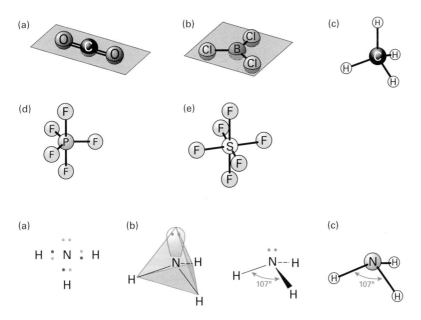

FIGURE 2.20 The structures of (a) CO_2, (b) BCl_3, (c) CH_4, (d) PF_5, and (e) SF_6.

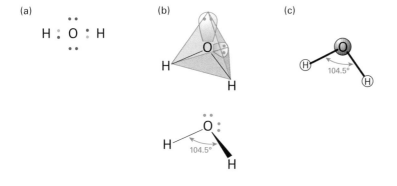

FIGURE 2.21 Ammonia NH_3. (a) The Lewis structure shows that the central atom has three bond pairs and one lone pair. (b) The arrangement of the electron pairs is distorted tetrahedral. (c) The shape of the ammonia molecule (i.e. the arrangement of the atoms) is pyramidal.

FIGURE 2.22 Water H_2O. (a) The Lewis structure shows that the central atom has two bond pairs and two lone pairs. (b) The arrangement of the electron pairs is distorted tetrahedral. (c) The molecule is bent into a broad V-shape.

the tetrahedral angle. You probably know the actual angles for ammonia (107°) and water (104.5°). It is most important to realize that it is **much too simplistic** to suggest that each extra lone pair reduces the angle by 2.5 degrees (see margin).

– Figure 2.23 shows that two lone pairs and four bond pairs, as is the case for xenon tetrafluoride (XeF_4), form a **square planar** shape. (The arrangement of the electron pairs is octahedral.)

– When the basic shape of the molecule is trigonal bipyramidal (five electron pairs), the effect of any lone pairs is more complicated (see Taking the Next Step 2.3).

• **We can also apply VSEPR to ions.** So, for example, the ammonium ion (NH_4^+) has exactly the same number of bond pairs (4) and lone pairs (0) as methane (CH_4), and has exactly the same shape (tetrahedral with bond angles of 109° 28′). We describe the two species as **isoelectronic** (having the same number of electrons).

The shape of the phosphine (PH_3) molecule is exactly the same as that for ammonia. However, the bond angle for PH_3 is 93° and not 107°. The shape of the hydrogen sulfide (H_2S) molecule is exactly the same as that for water. However, the bond angle for H_2S is 92° and not 104.5°.

FIGURE 2.23 The square-planar shape of xenon tetrafluoride, XeF_4.

PCl_4^+ (present in the solid form of phosphorus pentachloride PCl_5) is also tetrahedral, as is the case for the isoelectronic $SiCl_4$. The other ion present PCl_6^- is octahedral.

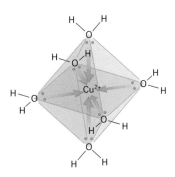

FIGURE 2.26 Boron trifluoride has an empty orbital in the valence shell. Ammonia donates an electron pair into this orbital to form a coordinate bond.

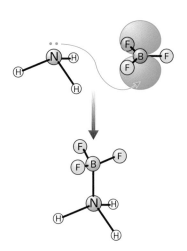

FIGURE 2.27 The shape of the dimer of aluminium chloride.

FIGURE 2.28 The complex ion $[Cu(H_2O)_6]^{2+}$. The oxygen atom in each water molecule donates an electron pair to the central metal ion. Each such electron pair constitutes a coordinate bond, shown by the blue arrow.

Another example of a coordinate bond often quoted is the ammonium ion: this is a poor example, because once the ion has formed it is impossible to work out which N–H bond is the coordinate bond: all the bonds are identical.

TAKING THE NEXT STEP 2.3
What about trigonal bipyramidal structures?

Unlike the tetrahedral and octahedral basic shapes, the trigonal bipyramidal basic shape does not have a unique bond angle (the angles within the middle triangle are 120° and the angle between the middle triangle and the top or bottom is 90°). We describe an electron pair within the middle triangle as **equatorial**; we describe an electron pair at top or bottom as **axial**.

We need to **consider the repulsions involving the *smallest* angle** to make the correct prediction of the observed shape. The molecule SF_4 has four bond pairs and one lone pair. The smallest angle here is 90°, so the lone pair must be placed equatorial to minimize LP/BP repulsions. (There are two LP/BP repulsions, compared with three if the lone pair is axial.) Figure 2.24 shows the resulting shape, which we call **see-saw**. (Because the lone pair is more repulsive than the bond pairs, the angle between the two equatorial bonds is slightly less than 120° and the angle between the axial and equatorial bonds is slightly less than 90°.)

The molecule ClF_3 has three bond pairs and two lone pairs. The smallest angle here is again 90°. Placing the two lone pairs axial would minimize the LP/LP repulsions but create six LP/BP repulsions at 90°. Placing the two lone pairs equatorial creates four LP/BP and two BP/BP repulsions at 90°, which is preferable. Figure 2.25 shows the resulting shape, which we call **T-shaped**. (Because the lone pairs are more repulsive than the bond pairs, the angle between the axial and equatorial bonds is slightly less than 90°.)

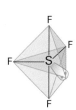

FIGURE 2.24 The see-saw shape of sulfur tetrafluoride, SF_4.

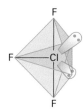

FIGURE 2.25 The T-shape of chlorine trifluoride, ClF_3.

2.5 COORDINATE BONDING

- In coordinate bonding, **one** of the two atoms sharing a pair of electrons donates **both** the electrons.

  - **Coordinate bonding** has two very common alternative names: **coordinate covalent bonding** and **dative covalent bonding**.

  - A good example occurs in the adduct $H_3N \rightarrow BF_3$ in which nitrogen donates its lone pair to boron, completing the latter's octet, as shown in Figure 2.26. The arrow indicates the coordinate bond.

  - The molecule $AlCl_3$ forms a dimer Al_2Cl_6 held together by coordinate bonds, as shown in Figure 2.27.

  - Coordinate bonding is central to the formation of **complexes** (Section 15.1), as shown in Figure 2.28.

2.6 IONIC BONDING

- **Ionic bonding** occurs when one atom **transfers an electron or electrons** to another atom, which creates oppositely charged ions; the **ions formed attract each other electrostatically, forming a crystal lattice**. Ionic bonding typically occurs between metals and non-metals.

- **The metal loses electrons to become a positive ion (cation).** Typically all the electrons in the valence shell are lost: removing electrons from an inner shell would require too much energy.

    - Sodium (Na) has one electron in the valence shell so forms the ion Na^+.
    - Magnesium (Mg) has two electrons in the valence shell so forms the ion Mg^{2+}.
    - Aluminium (Al) has three electrons in the valence shell so forms the ion Al^{3+}.

- **The non-metal gains electrons to become a negative ion (anion).** Typically the number of electrons added completes the valence shell of the non-metal.

    - Chlorine (Cl) has seven electrons in the valence shell so forms the ion Cl^-.
    - Oxygen (O) has six electrons in the valence shell so forms the ion O^{2-}.

- So the formula of sodium chloride is NaCl, magnesium chloride is $MgCl_2$, sodium oxide is Na_2O, and magnesium oxide is MgO. See Section 5.2 for analysis (using a Born–Haber cycle) of the enthalpy changes that occur during the formation of an ionic compound.

- A Deeper Look 2.2 describes how molecular orbital theory can even be applied to ionic bonding.

A DEEPER LOOK 2.2

Molecular orbital theory of ionic bonding

As the gap between the energy of the overlapping atomic orbitals widens, so the electron density becomes more unsymmetrical, which corresponds to more ionic character in the bond. If the energy gap between the atomic orbitals is *very* large, the bonding MO closely resembles the lower-energy atomic orbital. Transfer of an electron from the higher-energy atomic orbital is effectively an electron transfer from that atom to the other atom and hence creates a predominantly ionic bond.

We discuss the MO theory of polar covalent bonding in Taking the Next Step 2.2.

2

2.7 POLARIZATION OF IONIC BONDING

- An ionic bond will be distorted from the ***hypothetical*** complete transfer of electron density from one atom to the other because the electrostatic field created by the positive ion will attract the transferred electron to some extent. Figure 2.29 shows that the electron is *almost but not completely transferred* in sodium chloride. *Complete* electron transfer would result in exactly 10 electrons per Na^+ ion: the experimental value is 10.05.

- The magnitude of the electrostatic field depends on the charge density: the higher the charge density of the positive ion, the greater the polarization of the bond because of the stronger field created. The positive ion attracts electron density back towards itself, introducing partial covalent bonding. **The higher the charge density of the positive ion, the greater the covalent character in an ionic bond.**
  - The **charge density** equals the charge on the ion divided by its volume. The *greater* the charge on the ion, the larger its charge density. The *smaller* the size of the ion, the larger its charge density.

- Down a group, the charge on the ion usually remains the same while the size of the ion increases, hence **the charge density of the positive ion falls down a group** and there is less covalent character in the bond. In Group 2, beryllium compounds have the most covalent character.

- Across a period, two trends occur.
  - First, the charge on the positive ion increases as the number of valence electrons increases: Na^+, Mg^{2+}, Al^{3+}.
  - Second, these ions have the same number of electrons but increasing numbers of protons, so the ions get smaller.

- The combination of these two trends is that **the charge density of the ions increases from Na^+ to Mg^{2+} to Al^{3+}**, creating more covalent character for aluminium chloride than for sodium chloride.
  - $SiCl_4$ is much better described as having polar covalent bonding as the hypothetical Si^{4+} ion would be very strongly polarizing.

FIGURE 2.29 The experimental electron density for sodium chloride shows very little electron density between the sodium ions and the chloride ions. This picture supports the idea that electrons have been transferred from one atom to another to form ions. Diagrams like this are made by analysing the diffraction of X-ray beams by crystals of ionic compounds. Experimental measurement indicates 10.05 electrons per Na^+ ion: complete electron transfer would give a value of exactly 10.

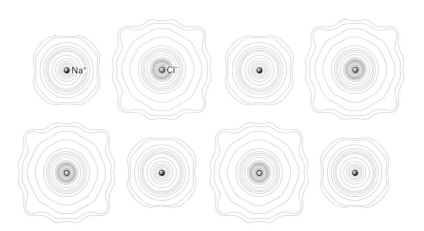

The other important factor affecting the extent of covalent character in a bond, which you are unlikely to have encountered pre-university, is the **polarizability (Section 4.2) of the negative ion**; see A Deeper Look 2.3.

A DEEPER LOOK 2.3

What effect does the negative ion have?

In general, **the larger the negative ion, the more polarizable it is**. So polarizability decreases from iodide to bromide to chloride to fluoride: silver iodide shows more covalent character than silver fluoride.

In general, **the larger the charge on the negative ion, the more polarizable it is**. So oxides will have more covalent character than the corresponding fluorides, despite fluoride ion F^- having the same electronic structure as oxide ion O^{2-}.

The combination of the greater charge density of the positive ion together with the greater polarizability of the negative ion means that for the isoelectronic compounds NaF to SiC:

- The ionic model is a good description of the bonding both for NaF and MgO, especially for NaF.

- The polar covalent model is a good description of the bonding both for AlN and SiC, especially for SiC.

MOLES

3.1 TYPES OF FORMULAE

Later we will mention structural and skeletal formulae (Section 16.1).

- There are different types of formulae in chemistry.

- The **empirical formula** (EF) is the simplest whole-number ratio of atoms of each element in a compound, see Section 11.5.

  Empirical formulae will rarely be discussed at university.

- The **molecular formula** (MF) is the actual number of atoms of each element in a compound.

  - The molecular formula is therefore a whole-number multiple of the empirical formula.

    ○ Ethene has an EF of CH_2 and an MF of C_2H_4.

    ○ Benzene has an EF of CH and an MF of C_6H_6.

    ○ Hydrogen peroxide has an EF of HO and an MF of H_2O_2.

Relative formula mass

- We can find the **relative formula mass** M_r of a compound by adding together all the relative atomic masses of the atoms present in the compound.

- We can use the term '**relative molecular mass**' equally well when the compound is simple molecular, but its use is discouraged when the compound is ionic.

3.2 THE MOLE AND THE AVOGADRO CONSTANT

- Chemists count atoms by weighing a large collection of them, in much the same way that bank clerks count ten-pence pieces in collections of 50 or sheets of paper are counted in reams (500 sheets). The principle is the same for the mole, only the numbers are much larger.

- We call the number of atoms in exactly 12 g of carbon-12 one mole.

- **One mole** (symbol: **mol**) is the amount of *any* substance that contains the same number of particles as there are atoms in exactly 12 g of carbon-12. The particles (atoms, molecules, ion pairs) must be specified.

> In 2018 IUPAC issued a new definition of one mole as the amount of substance containing exactly $6.02214076 \times 10^{23}$ particles. The new definition helpfully emphasizes that one mole is concerned with counting particles rather than measuring masses. However, this change is expected to make little difference in everyday usage.

- The number of atoms per mole is called the **Avogadro constant** (symbol: N_A or L).
- The Avogadro constant has the value 6.022×10^{23} mol^{-1}.

> Note very carefully that the Avogadro constant has a unit; it is *not* a pure number.

Although 'Avogadro constant' is its IUPAC name, you will also frequently see this called **Avogadro's constant**; a few people still use **Avogadro's number**.

3.3 MOLAR MASS AND AMOUNT IN MOLES

- The **molar mass** (symbol M or sometimes M_m) is the mass per mole of a substance. The molar mass has the same numerical value as the relative formula mass but it also has units of g mol^{-1}.
- The **amount of substance** (also called **amount in moles** or **chemical amount**), n, can be found using the formula:

  $$n = \text{mass/molar mass} = m/M$$

- An equation such as $Mg(s) + 2HCl(aq) \rightarrow MgCl_2(aq) + H_2(g)$ can be interpreted as follows: 1 mol Mg reacts with 2 mol HCl to form 1 mol MgCl$_2$ and 1 mol H$_2$. The numbers (here 1:2:1:1) are called the **mole ratios** (or **stoichiometric ratios**).
- So mass-to-mass calculations, which we discuss further below, simply use the mole ratios and the molar masses.
- The **percentage yield** = (actual mass of product/theoretical mass of product) × 100
  - We show two worked examples below.
- The **% atom economy** = (mass of desired product/total mass of reactants) × 100

> The yield in a reaction depends on the reactant that is present in the smallest amount: we call this the **limiting reactant**. See the percentage yield and theoretical yield calculations below for examples of its use.

The phrase 'number of moles' is frequently seen in place of 'amount in moles'. However, this phrase should be avoided, in much the same way as 'number of grams' is a poor substitute for 'mass' and 'number of kelvins' is a poor substitute for 'temperature'. Yet the phrase is still widely used, so do be aware of it.

The term 'limiting reactant' is not used very often pre-university.

3.4 STRATEGY FOR DOING MASS-TO-MASS CALCULATIONS

1 Use the given mass of substance A to find the amount in moles of A (by dividing by its molar mass).

2 Use the balanced chemical equation to state the mole ratio between substance B and A; then use that ratio to find the amount in moles of B.

3 Find the mass of B (by multiplying its amount in moles by its molar mass).

Example mass-to-mass calculation

What mass of sulfur is required to make 36.0 megatonnes of sulfuric acid?

1 First, find the amount in moles of sulfuric acid:

$$36.0 \text{ megatonnes} = 36.0 \times 10^6 \text{ tonnes} = 36.0 \times 10^{12} \text{ g}$$
$$M(H_2SO_4) = (2 \times 1.0 + 32.1 + 4 \times 16.0) \text{ g mol}^{-1} = 98.1 \text{ g mol}^{-1}$$
$$n(H_2SO_4) = 36.0 \times 10^{12} \text{ g}/98.1 \text{ g mol}^{-1} = 3.67 \times 10^{11} \text{ mol}$$

2 The balanced equation for the overall reaction is

$$2S + 2H_2O + 3O_2 \rightarrow 2H_2SO_4$$

Hence we need 1 mol S to form 1 mol H_2SO_4

$$n(S) = 3.67 \times 10^{11} \text{ mol}$$

3 Finally, find the mass of sulfur from its amount in moles:

$$m(S) = (3.67 \times 10^{11} \text{ mol}) \times (32.1 \text{ g mol}^{-1}) = 1.18 \times 10^{13} \text{ g}$$

We need 11.8 megatonnes of sulfur.

Example percentage yield calculation

When 10.0 g of ethanol is combusted in the lab in 4.00 g of oxygen, 1.33 g of water is produced. What is the percentage yield?

1 First find the amount in moles of each reactant:

$$M(CH_3CH_2OH) = (2 \times 12.0 + 6 \times 1.0 + 16.0) \text{ g mol}^{-1} = 46.0 \text{ g mol}^{-1}$$
$$n(CH_3CH_2OH) = 10.0 \text{ g}/46.0 \text{ g mol}^{-1} = 0.217 \text{ mol}$$
$$n(O_2) = 4.00 \text{ g}/32.0 \text{ g mol}^{-1} = 0.125 \text{ mol}$$

2 Use the balanced equation to work out which is the limiting reactant:

$$CH_3CH_2OH(l) + 3O_2(g) \rightarrow 2CO_2(g) + 3H_2O(l)$$

1 mol CH_3CH_2OH reacts with 3 mol O_2 so 0.217 mol CH_3CH_2OH reacts with

3×0.217 mol = 0.65 mol O_2 but there is only 0.125 mol O_2, hence oxygen is the limiting reactant.

3 3 mol O_2 reacts to form 3 mol H_2O, so 0.125 mol O_2 forms 0.125 mol H_2O and the theoretical mass of H_2O = (0.125 mol) × (18.0 g mol^{-1}) = 2.25 g

4 Hence percentage yield = (actual mass of H_2O/theoretical mass of H_2O) × 100

$$= (1.33 \text{ g}/2.25 \text{ g}) \times 100 = 59\% \text{ (2 sf)}.$$

3.5 THE IDEAL GAS EQUATION AND THE MOLAR VOLUME

- The **ideal gas model** (or **perfect gas model**, see Taking the Next Step 3.1) pictures a gas as a collection of hard spheres of negligible volume travelling in rapid random motion, with no forces between the spheres. All collisions are perfectly **elastic**: the spheres bounce off each other without their total kinetic energy changing. (The ideal gas model explains the three gas laws; see Taking the Next Step 3.2.)

- The resulting **ideal gas equation** is $pV = nRT$ where
  - p is the pressure exerted by the gas
  - V is the volume occupied by the gas
  - n is the amount in moles of the gas
  - R is the **gas constant** (or universal gas constant) ($8.314 \text{ J K}^{-1} \text{ mol}^{-1}$)
  - T is the thermodynamic temperature

  > Note which quantities are in capitals and which are in lower case. Sometimes you will see the pressure in capitals (possibly because p is being used for another quantity such as momentum) but IUPAC recommends use of lower case.

- When using the ideal gas equation, always use **SI units**:
  - p in Pa (pascal, N m^{-2} which can also be written as J m^{-3})
  - V in m^3
  - n in mol
  - T in kelvin

- Important interconversions are as follows:
  - $1 \text{ bar} = 100 \text{ kPa} = 1 \times 10^5 \text{ Pa}$
  - $1 \text{ atm (atmosphere)} = 1.013 \times 10^5 \text{ Pa}$
  - $1 \text{ m}^3 = 1000 \text{ dm}^3$, as shown in Figure 3.1
  - $1 \text{ dm}^3 = 1000 \text{ cm}^3$, as shown in Figure 3.2
  - $0 \text{ °C} = 273 \text{ K}$

  > An exceptionally important connection made at university is that the gas constant R is equal to the Avogadro constant N_A multiplied by the Boltzmann constant k (which interrelates the population of molecular energy levels):
  >
  > $$R = N_A k$$

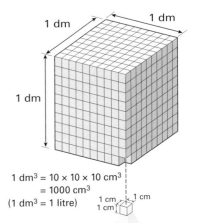

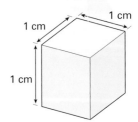

$1 \text{ dm}^3 = 10 \times 10 \times 10 \text{ cm}^3$
$= 1000 \text{ cm}^3$
($1 \text{ dm}^3 = 1 \text{ litre}$)

FIGURE 3.2 The relationship between the cubic decimetre (dm³) and the cubic centimetre (cm³). You may also come across 'cc' used for cubic centimetre in older textbooks.

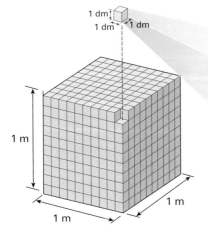

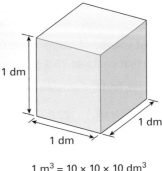

$1 \text{ m}^3 = 10 \times 10 \times 10 \text{ dm}^3$
$= 1000 \text{ dm}^3$

FIGURE 3.1 The relationship between the cubic metre (m³) and the cubic decimetre (dm³).

3

Why is 'perfect gas' preferred to 'ideal gas'?

At university you will study ideal solutions. **Ideal solutions** form when the forces between the molecules in two liquids are the same: ideal solutions obey Raoult's law (Section 17.1). By analogy, an ideal gas would imply that the forces between the molecules in a gas are the same *but not necessarily zero*. Because the ideal gas model assumes that there are *no* forces, it is better to use the term '**perfect gas**', as is common at university.

* We can use the ideal gas equation to calculate the molar volume of an ideal gas under any conditions of temperature and pressure.

* The **molar volume** of a gas (symbol V_m) is the volume occupied per mole.

  – At a temperature of 20 °C and a pressure of 1.00 bar, the ideal gas equation predicts that

$$V_m = V/n = RT/p = (8.314\,\text{J}\,\text{K}^{-1}\,\text{mol}^{-1})(293\,\text{K})/(1.00 \times 10^5\,\text{J}\,\text{m}^{-3})$$
$$= 0.0244\,\text{m}^3\,\text{mol}^{-1}$$

* In the more convenient unit of $dm^3\,mol^{-1}$, and retaining only 2 sf, the **molar volume of an ideal gas at 20 °C and 1 bar is 24 dm³ mol⁻¹**, familiar from earlier courses.

* Figure 3.3 shows how close the actual values for a number of gases are to the predicted value.

* We can find the amount in moles of a gas as follows:

  – **n = volume/molar volume = V/V_m**

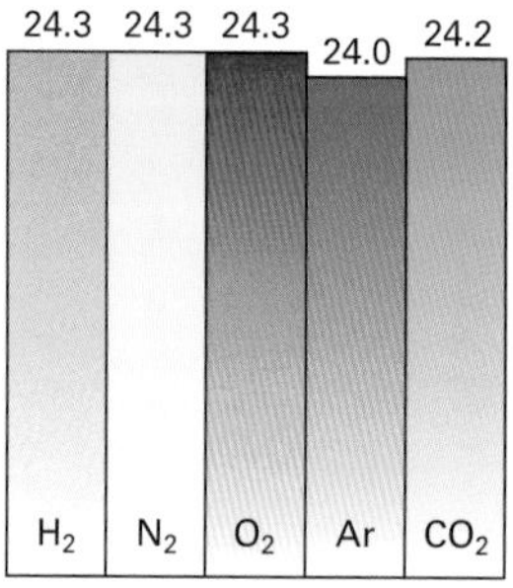

FIGURE 3.3 The molar volumes of different gases are all very similar. Figures are given in dm³ mol⁻¹ at 20 °C and 1 bar.

Bar chart values: H₂ 24.3, N₂ 24.3, O₂ 24.3, Ar 24.0, CO₂ 24.2

3.6 STRATEGY FOR DOING MASS-TO-VOLUME CALCULATIONS

This follows the strategy for mass-to-mass calculations set out above with the only change being to the final step (3), which becomes:

3 Find the volume of B (by multiplying its amount in moles by its molar volume).

Example mass-to-volume calculation

What volume of nitrogen gas (at 20 °C and 1 bar) is produced from an air-bag that contains 70.3 g of sodium azide NaN_3?

1 First, find the amount in moles of sodium azide:

$$M(NaN_3) = (23.0 + 3 \times 14.0)\,\text{g}\,\text{mol}^{-1} = 65.0\,\text{g}\,\text{mol}^{-1}$$
$$n(NaN_3) = 70.3\,\text{g}/65.0\,\text{g}\,\text{mol}^{-1} = 1.08\,\text{mol}$$

2 Sodium azide decomposes as follows:

$$2NaN_3(s) \rightarrow 2Na(s) + 3N_2(g)$$

Hence $1.08\,mol$ NaN_3 produces $\left(\dfrac{3}{2}\right) \times 1.08\,mol = 1.62\,mol\,N_2$.

3 Finally, find the volume of nitrogen gas from its amount in moles:

$$V(N_2) = (1.62\,mol) \times (24.4\,dm^3\,mol^{-1}) = 40\,dm^3\ (2\ sf).$$

We calculated the molar volume of an ideal gas at 20 °C and 1 bar just above.

Airbags are designed to inflate in about 20 milliseconds.

Strategy for doing volume-to-volume calculations for gases

Volume-to-volume calculations become trivial because the molar volume of all gases is almost the same. Hence **Gay-Lussac's law** states that the volumes of gases that react (and the volumes of the products if gaseous) are proportional to their mole ratios.

TAKING THE NEXT STEP 3.2

The three gas laws

We can use the ideal gas equation $pV = nRT$ to explain the three laws which were found several centuries ago to describe the behaviour of gases, the first two of which were based on experiments.

Boyle's law (1662), see Figure 3.4.

Boyle's law states that the volume of a fixed mass of gas (at constant temperature) is inversely proportional to its pressure.

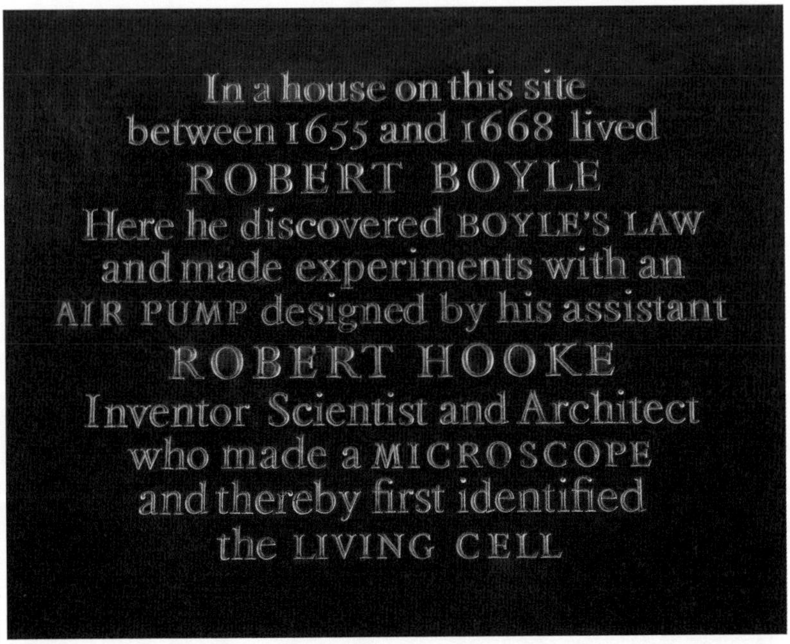

In a house on this site
between 1655 and 1668 lived
ROBERT BOYLE
Here he discovered BOYLE'S LAW
and made experiments with an
AIR PUMP designed by his assistant
ROBERT HOOKE
Inventor Scientist and Architect
who made a MICROSCOPE
and thereby first identified
the LIVING CELL

FIGURE 3.4 A plaque commemorating Robert Boyle and Robert Hooke in High Street, Oxford.

A fixed mass of gas means that n is constant; a constant temperature ensures that T is constant. The gas constant R is a universal constant, so under the stated conditions nRT is constant. So given that $pV = nRT$, pV must be constant; hence V is inversely proportional to p.

Charles's law (1787), see Figure 3.5.

Charles's law states that the volume of a fixed mass of gas (at constant pressure) is proportional to its thermodynamic temperature.

Rearrange the ideal gas equation to make V the subject: $V = nRT/p$.

A fixed mass of gas means that n is constant; a constant pressure ensures that p is constant. The gas constant R is a universal constant, so under the stated conditions nR/p is constant. So given that $V = nRT/p$, V is directly proportional to T.

Avogadro's principle (1811), see Figure 3.6.

Avogadro's principle states that equal volumes of different gases at the same temperature and pressure contain the same number of molecules.

FIGURE 3.5 A contemporary illustration of the first flight by Jacques Charles with Nicolas-Louis Robert, December 1, 1783. Viewed from the Place de la Concorde to the Tuileries Palace. Later in the day, Charles ascended again alone to a world-record height of about 2.7 km. A hydrogen-filled balloon was called a Charlière in his honour.

FIGURE 3.6 An Italian stamp depicting Avogadro's principle.

Rearrange the ideal gas equation to make V the subject: $V = nRT/p$.

A constant temperature ensures that T is constant; a constant pressure ensures that p is constant. The gas constant R is a universal constant, so under the stated conditions RT/p is constant. So given that $V = nRT/p$, V is directly proportional to n.

> Although the ideal gas model is commonly used in chemistry, be aware that it models a simplified system and not reality. During your time at university you will encounter other models which describe real systems even better (at the expense of being more complicated).

3.7 THE CONCENTRATION OF SOLUTIONS

- A **solution** consists of a **solute** dissolved in a **solvent**.

- We can express the concentration of the solution in a number of ways, including the **mass concentration**, which specifies the mass of the solute dissolved per cubic decimetre (dm^3) of the solution.

- In chemistry the most usual measure is the **molar concentration**, often shortened simply to 'concentration': the amount in moles of the solute dissolved per cubic decimetre (dm^3) of the *solution*.

    concentration = amount in moles of solute/volume of solution

    or in symbols $c = n/V$

- To find the amount in moles, rearrange this equation to give

    $n = Vc$

    - It is particularly important to remember that the volume will usually be given in cm^3 and so we need to convert to dm^3, as shown in Figure 3.2, before multiplying by the concentration in $mol\ dm^{-3}$:

    $1000\ cm^3 = 1\ dm^3$

Example theoretical yield calculation

What volume of hydrogen gas (at 20 °C and 1 bar) is produced from 500 cm^3 of 2.00 mol dm^{-3} hydrochloric acid and excess zinc?

1 First, find the amount in moles of the acid added:

$$n(HCl) = V(HCl) \times c(HCl) = \left(\frac{500}{1000} \right) dm^3 \times 2.00\, mol\, dm^{-3} = 1.00\, mol$$

2 Zinc reacts with hydrochloric acid according to the following equation:

$$Zn(s) + 2HCl(aq) \rightarrow ZnCl_2(aq) + H_2(g)$$

$$n(H_2) = \frac{1}{2} n(HCl) = \frac{1}{2} \times 1.00\, mol = 0.500\, mol$$

3 Finally, find the volume of hydrogen gas from its amount in moles:

$$V(H_2) = (0.500\, mol) \times (24.4\, dm^3\, mol^{-1}) = 12\, dm^3 \; (2\, sf)$$

Section 3.5 showed how to calculate the molar volume of an ideal gas at 20 °C and 1 bar.

Comment: In this reaction the limiting reactant (Section 3.3) is the acid: the zinc is in excess.

A Deeper Look 3.1 shows a harder yield calculation.

Example titration calculation

We can use **titrations** (Section 7.4) to measure the volume of one solution needed to neutralize a given volume of another solution.

The NaOH will have been measured using a **pipette**, and the HCl added to this using a **burette**. Figure 3.7 illustrates the use of a burette.

What is the concentration of NaOH(aq) if 25.0 cm^3 of the solution exactly neutralizes 40.0 cm^3 of 0.250 mol dm^{-3} HCl(aq)?

1 First, find the amount in moles of the acid added:

$$n(HCl) = \left(\frac{40.0}{1000} \right) dm^3 \times 0.250\, mol\, dm^{-3}$$

$$= 1.00 \times 10^{-2}\, mol$$

2 Sodium hydroxide reacts with hydrochloric acid according to the following equation:

$$NaOH(aq) + HCl(aq) \rightarrow NaCl(aq) + H_2O(l)$$

NaOH and HCl react in a 1:1 mole ratio hence
$$n(NaOH) = 1.00 \times 10^{-2}\, mol$$

3 Finally, find the concentration from the amount in moles divided by the volume:

$$c(NaOH) = n(NaOH)/V(NaOH) = 1.00 \times 10^{-2}\, mol / \left(\frac{25.0}{1000} \right) dm^3$$

$$= 0.400\, mol\, dm^{-3}$$

FIGURE 3.7 A burette is used to deliver a solution carefully.

3

A DEEPER LOOK 3.1

What about a combined example?

A scientist working in the lab at 20 °C on a nice spring day bubbles chlorine gas through a solution of sodium hydroxide, hoping to make sodium hypochlorite (NaClO, Section 13.6), sodium chloride, and water. Knowing that the pressure of chlorine is 0.50 bar, the concentration of sodium hydroxide is 0.10 g dm^{-3}, and the volume of the reactor tank is 300 cm^3, predict the theoretical mass of sodium hypochlorite formed. The scientist produces only 0.010 g of sodium hypochlorite, so what is the percentage yield?

Answer (see Section 3.4 for an example of a percentage yield calculation):

1 First find the amount in moles of each reactant:

> **For chlorine**, there is 0.50 bar in 300 cm^3. Using the ideal gas equation,
>
> $$n = \frac{pV}{RT}$$
>
> In S.I. units: 0.50 bar = 0.50 × 10^5 Pa = 0.50 × 10^5 J m^{-3} and 300 cm^3 = 300 × 10^{-6} m^3
>
> $$n(Cl_2) = \frac{(0.50 \times 10^5 \text{ J m}^{-3}) \times (300 \times 10^{-6} \text{ m}^3)}{(8.314 \text{ J K}^{-1} \text{ mol}^{-1}) \times (293 \text{ K})} = 6.2 \times 10^{-3} \text{ mol}$$
>
> **For sodium hydroxide**, there is 0.10 g dm^{-3} in 300 cm^3 (300 × 10^{-3} dm^3) hence a mass of (0.10 g dm^{-3}) × (300 × 10^{-3} dm^3) = 0.030 g.
>
> To find the amount in moles, use $n = m/M$
>
> $M(\text{NaOH}) = (23.0 + 16.0 + 1.0) = 40.0 \text{ g mol}^{-1}$
>
> $n(\text{NaOH}) = 0.030 \text{ g}/40.0 \text{ g mol}^{-1} = 7.5 \times 10^{-4} \text{ mol}$

2 Use the balanced equation (Section 13.6) to work out which is the limiting reactant:

> $$Cl_2(g) + 2NaOH(aq) \rightarrow NaClO(aq) + NaCl(aq) + H_2O(l)$$

2 mol NaOH reacts with 1 mol Cl_2 so 7.5×10^{-4} mol NaOH reacts with 3.75×10^{-4} mol Cl_2, but there is 6.2×10^{-3} mol Cl_2, so NaOH is the limiting reactant.

3   2 mol NaOH reacts to form 1 mol NaClO, so 7.5×10^{-4} mol NaOH forms 3.75×10^{-4} mol NaClO.

$M(NaClO) = (23.0 + 35.5 + 16.0)$ g mol$^{-1} = 74.5$ g mol^{-1},

hence the theoretical mass of NaClO $= (3.75 \times 10^{-4}$ mol$) \times (74.5$ g mol$^{-1})$

$= 0.0279$ g.

4   Hence percentage yield $=$ (actual mass of NaClO/theoretical mass

of NaClO$) \times 100$

$= (0.010$ g$/0.0279$ g$) \times 100$

$= 36\%$ (2 sf).

STATES OF MATTER

4.1 POLAR BONDS AND POLAR MOLECULES

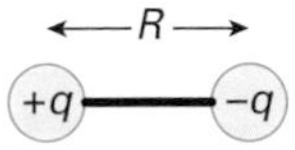

FIGURE 4.1 A dipole consists of a positive charge separated from a negative charge.

- A **polar molecule** is one that has a dipole moment.

- A **dipole** exists if a positive charge $+q$ is separated from a negative charge $-q$ by a distance R: this produces a **dipole moment** μ of magnitude qR, as shown in Figure 4.1.

- To work out whether **polyatomic molecules** (molecules containing more than two atoms) have a dipole moment, decide first whether each bond is polar (Section 2.2) and then whether the shape of the molecule means that the bond dipoles cancel out.

- Some molecules have no dipole moment:

  - All **homonuclear diatomic molecules** (molecules containing two identical atoms), such as H_2, N_2 or O_2, have no dipole as the bond is not polar.

  - **Symmetrical molecules** such as CO_2, BCl_3, CH_4, PF_5, and SF_6 have bonds that *are* polar (Section 2.2), but the *bond dipoles cancel out*, producing no dipole moment, as shown in Figure 4.2.

- Other molecules have a non-zero dipole moment:

  - All **heteronuclear diatomic molecules** (molecules containing two non-identical atoms), such as HF, HCl, and HBr, have a dipole moment because of the polar covalent bond (Section 2.2) between the two atoms.

  - **Unsymmetrical molecules**, such as NH_3 or H_2O (as shown in Figure 4.3), have bond dipoles which do not cancel out (see A Deeper Look 4.1), leaving the molecule as a whole with a dipole moment.

(a)
O C O
(b)
Cl B Cl Cl
(c)
H C H H H
(d)
F F P F F F
(e)
F F F S F F F

FIGURE 4.2 The molecules CO_2, BCl_3, CH_4, PF_5, and SF_6 have zero dipole moment.

(a)
N H H H
N H H H
107°
(b)
O H H
O H H
104.5°

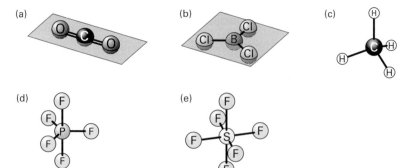

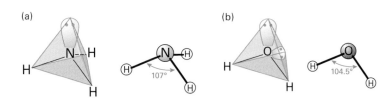

FIGURE 4.3 (a) Ammonia and (b) water have dipole moments.

A DEEPER LOOK 4.1

Dipoles as vectors

We can treat each bond dipole as a vector (pointing from negative to positive) and we can then add the various vectors together, as shown in Figure 4.4, to find out whether the molecule has a net dipole moment or not. Figure 4.4 (b) shows that exact cancellation occurs on vector addition. In (c) there is a resultant dipole. In (d) the bond dipoles point more closely in a similar direction so the resultant dipole is larger.

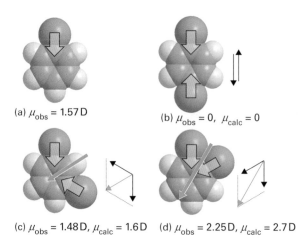

(a) μ_{obs} = 1.57 D
(b) μ_{obs} = 0, μ_{calc} = 0
(c) μ_{obs} = 1.48 D, μ_{calc} = 1.6 D
(d) μ_{obs} = 2.25 D, μ_{calc} = 2.7 D

FIGURE 4.4 Combining bond dipoles. In (b) the two bond dipoles cancel exactly. There is a resultant dipole in both (c) and (d): the resultant dipole in (d) is larger.

4.2 INTERMOLECULAR FORCES

- **Intermolecular forces** are the forces between molecules.
  - We can classify intermolecular forces as dipole-dipole forces, dispersion forces, and hydrogen bonding.
- **Dipole-dipole forces** occur when a polar molecule interacts with another polar molecule, see Figure 4.5.
- **Dispersion forces** occur between *all* molecules (and even atoms). Taking the Next Step 4.1 explains the origin of dispersion forces.
  - Dispersion forces are also commonly called induced dipole-induced dipole forces or **London forces** (after Fritz London).
  - Dispersion forces get larger as the polarizability of the molecule increases (see Taking the Next Step 4.2). In simple terms, the polarizability of the molecule is proportional to the number of electrons present, so *dispersion forces increase as the number of electrons in a molecule increases*.

  You may also have come across permanent dipole-induced dipole forces (**Debye forces**, after Peter Debye). These are rarely important quantitatively, contributing less than 0.1% as much as the dispersion force for CO and only 3% as much as the dispersion force for HCl. The Debye force's largest contribution is for NH_3, where it makes up 5% of the total intermolecular force.

Dipole-dipole forces are also commonly called permanent dipole-permanent dipole forces and very occasionally **Keesom forces** (after Willem Keesom).

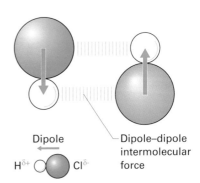

FIGURE 4.5 Dipole-dipole forces.

A common misconception is that the term **van der Waals forces** can be used synonymously with dispersion forces. Intermolecular forces in general **can** be called van der Waals forces. (Johannes van der Waals was the first person to identify that molecules in gases can attract each other.) Dispersion forces are only one contributor to the overall intermolecular forces. It is very significant that van der Waals died seven years *before* the dispersion force was discovered!

TAKING THE NEXT STEP 4.1

How do dispersion forces arise?

An *instantaneous* dipole can form in one molecule when the electron density fluctuates faster than the nuclei can move. This transient dipole can induce an instantaneous dipole in another molecule. As the first transient dipole varies, the second induced dipole will follow it, so the interaction is always attractive. The two dipoles get into step, as shown in Figure 4.6.

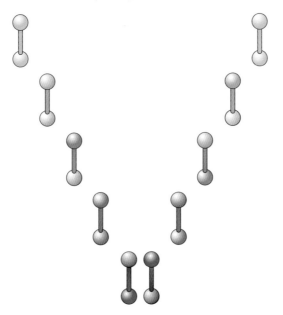

FIGURE 4.6 A schematic diagram explaining the origin of dispersion forces. When two non-polar molecules (such as two chlorine molecules) approach each other, instantaneous dipoles form that *get into step*. The opposite partial charges on the molecules then cause attraction. As the molecules approach closer, the instantaneous dipoles become larger in magnitude (shown by the greater intensity of colouring). The force is always an attraction but the direction of the dipole on an individual molecule is random; hence the figure shows the direction switching twice.

> ## TAKING THE NEXT STEP 4.2
> ### What is polarizability?
>
> The **polarizability** measures the ease with which a molecule's electron density can be distorted. The size of the induced dipole moment μ created by an electric field is directly proportional to the electric field strength E, with the constant of proportionality being the polarizability α of the molecule: $\mu = \alpha E$. The greater the polarizability of the molecule, the larger the induced dipole moment.

* **Hydrogen bonding** occurs when a hydrogen atom covalently bonded to a *small, highly electronegative* atom (F, O or N) is attracted to the lone pair of another F, O or N atom.

  > IUPAC has recently broadened this definition significantly to include *all* elements more electronegative than hydrogen (including carbon!). It remains to be seen whether this change will find favour within the chemistry community.

  - The bond between H and O (for example) is highly polar with a large partial positive charge on the hydrogen atom. This is then attracted to the lone pair on O.
  - Hydrogen bonding is the strongest intermolecular force; see the discussion on simple molecular solids below.
  - Hydrogen bonding plays a crucial role holding together the base pairs in the DNA molecule, as shown in Figure 4.7, and providing the secondary structure of proteins (Section 24.6).
  - Hydrogen bonding is also responsible for the low density of ice compared to liquid water (see Figure 4.11).

FIGURE 4.7 (a) Three hydrogen bonds hold together the bases cytosine and guanine. (b) Two hydrogen bonds hold together the bases thymine and adenine. The orange dashed lines indicate the hydrogen bonds.

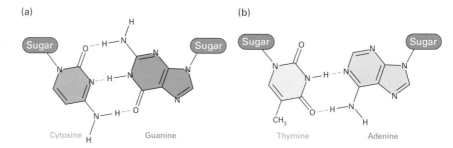

4.3 SOLIDS, LIQUIDS, AND GASES

* The particles in **solids** are close together, held in fixed positions about which they vibrate.

  - The particles in **crystalline solids** have a definite, regular structure.

* The particles in **liquids** are further apart than in solids but much closer than in gases; they diffuse rather than vibrate.

* The particles in **gases** are widely spaced and are in rapid, random motion.

Melting and boiling points

- At the **melting point** of a solid, the energy supplied is sufficient to *weaken* the attractive forces, listed in Table 4.1, between the particles.
- At the **boiling point** of a liquid, the energy supplied is sufficient to *break* the attractive forces between the particles.

TABLE 4.1 The typical magnitude of attractive forces.

Attractive force	Typical magnitude/kJ mol^{-1}
Covalent bonding	350
Dipole-dipole force	2
Dispersion force	5
Hydrogen bonding	20

4.4 TYPES OF CRYSTAL

There are four main types of crystalline solids: simple molecular, giant covalent, ionic, and metallic.

- **Simple molecular solids** are held together by intermolecular forces:
  - Intermolecular forces are relatively weak, so the melting and boiling points of simple molecular solids are generally low.
  - Very small molecules such as H_2, N_2, and O_2 are gases at room temperature.
  - Molecules with stronger intermolecular forces may be liquids or even solids.
    - For the halogens, dispersion forces increase as the number of electrons increases: at room temperature, F_2 and Cl_2 are gases, Br_2 is a liquid, and I_2 is a solid, as shown in Figure 4.8.
  - For the alkanes the dispersion forces increase as the number of carbon atoms increases, so the first four alkanes are gases, then come liquids, and eventually solids. Figure 4.9 shows the increasing boiling points of the alkanes.
  - The trend of increasing boiling points as the size of the molecule increases is disrupted by the existence of hydrogen bonding for ammonia, water, and hydrogen fluoride (which makes their boiling points anomalously high), as shown in Figure 4.10.
  - The presence of hydrogen bonding dominates the structure of ice, see Figure 4.11.
  - Simple molecular solids also have the following characteristics:
    - They tend to be more soluble in organic solvents rather than in water due to the dispersion forces these organic solvents provide.
    - They are non-conductors, as there are no charged particles present.

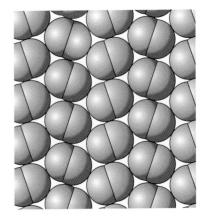

FIGURE 4.8 The structure of solid iodine.

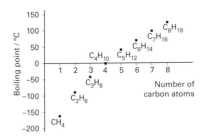

FIGURE 4.9 The boiling points of the alkanes increase as the dispersion forces increase.

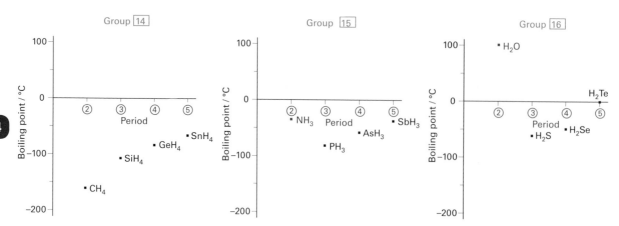

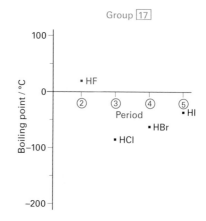

FIGURE 4.10 The boiling points increase steadily for the hydrides of Group 14 as dispersion forces increase down the group but hydrogen bonding in ammonia, water, and hydrogen fluoride make their boiling points anomalously high.

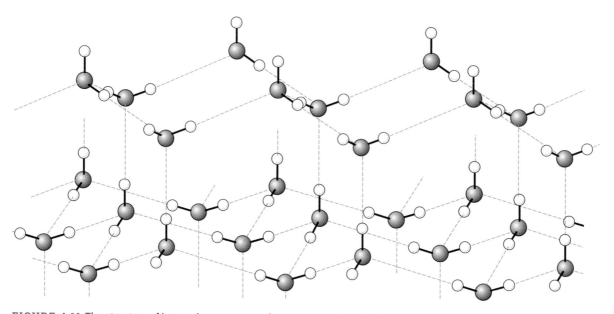

FIGURE 4.11 The structure of ice: each oxygen atom forms two covalent bonds and two hydrogen bonds (shown by the orange dashed lines).

- **Giant covalent solids** are held together by actual covalent bonds:
  - Covalent bonds are strong, as seen in Table 4.1, so the melting and boiling points of giant covalent solids are very high.
  - Diamond and graphite are giant covalent solids: diamond's melting point is above 3500 °C. Figure 4.12 shows the structures of diamond and graphite.
  - In addition to having very high melting and boiling points, giant covalent solids have the following characteristics:
    - They are non-conductors (with the exception of graphite, Section 2.3).
    - They are insoluble both in water and in organic solvents.

- **Ionic solids** are held together by the electrostatic attraction between oppositely charged ions:
  - Electrostatic attractions between oppositely charged ions are strong, so the melting and boiling points of ionic solids are generally much higher than those of simple molecular solids.
  - Sodium chloride is a typical ionic solid.

Figure 4.13 shows the crystal structure of sodium chloride (**rock salt**).

The phrase 'giant ionic solid' is tautologous: *all* ionic solids have a giant structure.

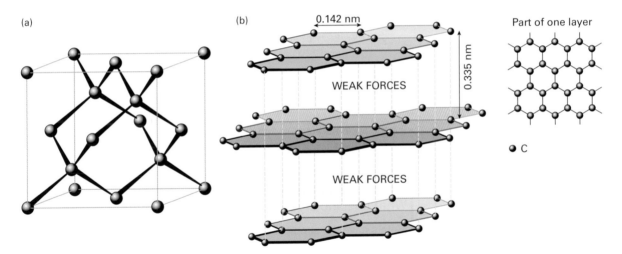

FIGURE 4.12 The structures of (a) diamond and (b) graphite.

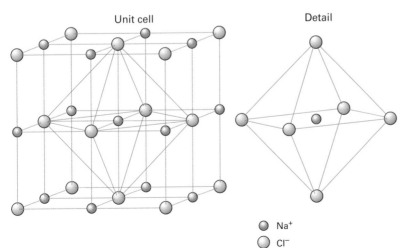

FIGURE 4.13 The **rock-salt structure** of sodium chloride: there are six chloride ions around each sodium ion in an octahedral arrangement. A **unit cell** is the small 3D figure from which we can construct the whole crystal by replication and translation in every direction.

- In addition to having high melting and boiling points, ionic solids have the following characteristics:
  - ○ They are usually soluble in water, as the ions are attracted to the dipole of water (Section 4.1).
  - ○ When molten or in solution, they conduct electricity as the *ions are free to move*.
- **Metallic solids** are held together by metallic bonding.
- **Metallic bonding** occurs when each atom contributes its valence electrons into a *delocalized 'sea' of electrons* which holds the resulting metal *ions* together.
  - The magnitude of metallic bonding varies very considerably from one metal to another, so no general statement about the melting and boiling points is appropriate. However, only mercury is a liquid at room temperature.
  - Metallic solids have the following characteristics:
    - ○ They conduct electricity both when solid and when molten.
    - ○ They are malleable (can be hammered) and ductile (can be drawn into wires) as the delocalized sea of electrons allows the metal ions to move.
    - ○ They are insoluble in water but several of them will *react* with water (Section 12.3).
  - The majority of metals (but by no means all) are close-packed, with twelve nearest neighbours.

A Deeper Look 4.2 gives more details on common crystal structures, including those of magnesium and calcium.

A DEEPER LOOK 4.2

What specific crystal structures do we commonly see?

Metals have the simplest crystal structures, as there is only one type and size of ion to fit together. There are in fact *two* close-packed structures that we can build: **hexagonal close-packed (hcp)**, as shown in Figure 4.14, and **cubic close-packed (ccp)**, as shown in Figure 4.15.

The ions in layer 2 lie in the hollows between the ions in layer 1   The ions in layer 3 lie directly above the ions in layer 1   Space-filling model of unit cell   Ball-and-stick model of unit cell

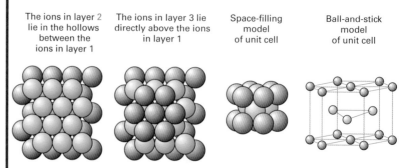

FIGURE 4.14 The **hexagonal close-packed (hcp)** structure: the third layer lies directly above the first, giving a pattern we can describe as ABABAB.

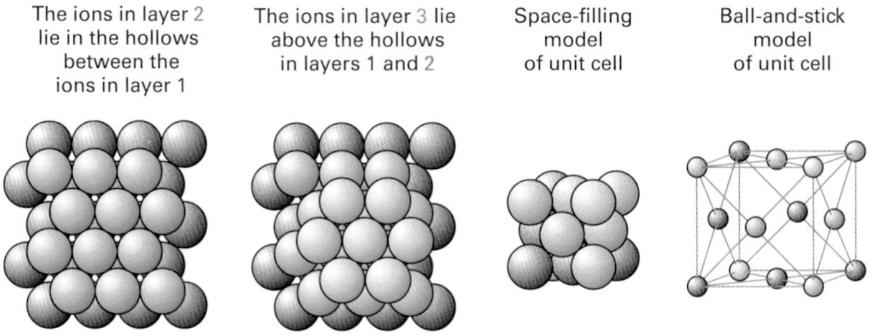

The ions in layer 2 lie in the hollows between the ions in layer 1

The ions in layer 3 lie above the hollows in layers 1 and 2

Space-filling model of unit cell

Ball-and-stick model of unit cell

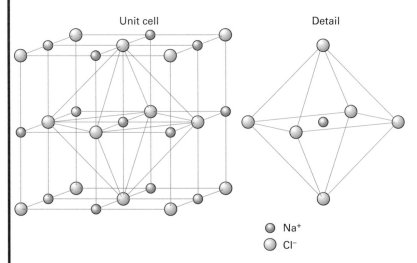

FIGURE 4.15 The **cubic close-packed (ccp)** structure: the third layer differs from both of the first two layers, giving a pattern we can describe as ABCABC. The unit cells use the same colouring scheme for the close-packed layers as in the second diagram, but the layers are rotated so that they are diagonal.

The structure adopted by magnesium is hcp (see Figure 4.14), whereas the structure adopted by calcium is ccp (see Figure 4.15). Rotating the ccp layers so that they are diagonal, as shown in Figure 4.15, gives a unit cell with *a ball at each corner of a cube* together with *another ball in the centre of each face of the cube*. This explains the very common alternative name for the ccp structure: **face-centred cubic (fcc)**.

The most important crystal structure for compounds with a 1:1 ratio of elements is the rock-salt structure (Figure 4.13): for example, seventeen Group 1 metal halides, silver chloride and bromide, together with the oxides of magnesium, calcium, and iron(II) all crystallize with this structure. Figure 4.16 shows the rock-salt structure again: we can interpret it as two interpenetrating face-centred cubic arrays—one of sodium ions and the other of chloride ions.

An alternative way of interpreting this structure imagines the smaller sodium ions fitting into holes within a face-centred cubic array of chloride ions. Each of these sodium ions is surrounded by six chloride ions in an octahedral arrangement (Figure 4.13) and so we call these holes **octahedral holes**, as shown in the detail in Figure 4.16.

Unit cell   Detail

○ Na+
○ Cl−

FIGURE 4.16 The **rock-salt structure** of sodium chloride: the **coordination number** (the number of nearest ions of the opposite charge) of each ion is 6. The blue lines outline one octahedral hole.

A face-centred cubic array has another set of holes, called **tetrahedral holes**. Filling *all* of them gives an important crystal structure for compounds with a 2:1 ratio of elements: the **antifluorite structure**, shown in Figure 4.17, adopted by the oxides of lithium, sodium, potassium, and rubidium, for example. Again, the metal ions are in the holes in a face-centred cubic array of oxide ions. Swapping the positive and negative ions around generates the **fluorite structure** itself, named after the mineral fluorite (calcium fluoride, CaF_2).

Finally, filling only *half* of the tetrahedral holes in a regular fashion gives the **zinc-blende structure** adopted by one solid form of zinc sulfide. This is another important structure for compounds with a 1:1 ratio of elements (such as the widely used semiconductor gallium arsenide GaAs, Figure 2.12). Figure 4.18 shows

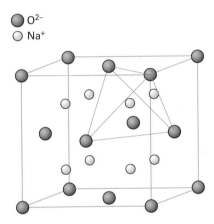

FIGURE 4.17 The **antifluorite structure** adopted by sodium oxide, Na_2O. There are 8 sodium ions around each oxide ion and 4 oxide ions round each sodium ion. The blue lines outline one tetrahedral hole.

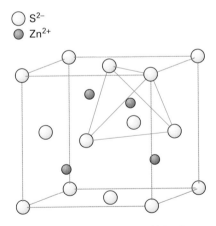

FIGURE 4.18 The **zinc-blende structure** of zinc sulfide: the coordination number of each ion is 4. The blue lines outline one tetrahedral hole. If the zinc and sulfur ions were identical, the structure would be the same as the diamond structure, which is redrawn in Figure 4.19.

the zinc-blende structure. (Figure 4.19 shows the structure of diamond, which is the same structure as in Figure 4.18, but now with both the yellow and brown spheres being the same.)

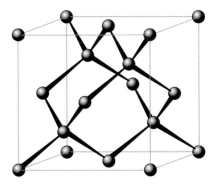

FIGURE 4.19 The diamond structure. Compare with Figure 4.18.

THERMOCHEMISTRY

5.1 ENTHALPY CHANGE

- The **enthalpy change ΔH** is the **heat added to a system at constant pressure**. A Deeper Look 5.1 explains this in more detail.
  - If heat is taken in, ΔH is positive and the reaction is **endothermic**.

    Technically, this is called the **acquisitive convention**.
  - If heat is given out, ΔH is negative and the reaction is **exothermic**.

    You may have seen the enthalpy change defined as the heat (energy) *change* at constant pressure. There is an obvious problem with this: it gives no information on the *sign* of the change. Taking the Next Step 5.1 discusses another serious criticism.

If you take money out of your bank account, your bank balance goes down.

TAKING THE NEXT STEP 5.1

Why is 'heat change' incorrect?

Hess's law (Section 5.2) states that the standard enthalpy change is independent of the route between reactants and products. Enthalpy is therefore a **state function**, one that depends only on the system's current state and is independent of the path by which that state is reached. Other examples of state functions are internal energy (A Deeper Look 5.1), entropy (Section 9.1), and Gibbs energy (Section 9.3).

In contrast, heat is *not* a state function, as we can generate any particular internal energy change (A Deeper Look 5.1) by any combination of heat and/or work. We can produce an internal energy increase of 100 kJ by adding 100 kJ of heat and 0 kJ of work or 50 kJ of heat and 50 kJ of work through to 0 kJ of heat and 100 kJ of work. Heat and work are **path functions**, which depend on the route taken from the initial state to the current state.

A simple analogy is that climbing the Eiger involves the same altitude increase whatever route you take from Kleine Scheidegg to the summit. However, the difficulty of reaching the summit is much greater for the north wall (Eigerwand, first conquered in 1938, illustrated in Figure 5.1) than for the west flank (first successfully climbed in 1858).

FIGURE 5.1 Eigerwand, the north wall of the Eiger, seen from Kleine Scheidegg.

A DEEPER LOOK 5.1

Why is enthalpy defined in this way?

The First Law of thermodynamics (Section 5.2) explains why the increase in **internal energy** of a system is exactly equal to the energy added by **heating** the system plus the energy added by **doing work** on the system. No energy is lost. In symbols,

$$\Delta U = q + w \tag{1}$$

where ΔU is the internal energy change, q is the heat added **to** the system, and w is the work done **on** the system. If we add 60 kJ by heating and 40 kJ by doing work, the internal energy increases by 100 kJ: $\Delta U = 100$ kJ.

At the molecular level, doing work causes uniform motion whereas heating causes random motion.

Chemical reactions often produce gases, and any gas evolved would have to **do** work to push against the external pressure. We can use Figure 5.2 to find the work done in expansion.

> work done = force × distance moved in the direction of the force
> pressure = force/area so force = pressure × area
> work done **by** the system = pressure × area × distance moved
> = pressure × volume increase
> work done **on** the system = −work done **by** the system
> In symbols, $w = -p\Delta V$ $\tag{2}$

Substituting equation (2) into equation (1) we find

$$\Delta U = q - p\Delta V \tag{3}$$

This means that in any chemical reaction involving gases, it would be necessary to measure the volume of gas evolved and the external pressure. This would be a rather annoying extra task, so enthalpy is defined precisely to avoid the need for making this measurement.

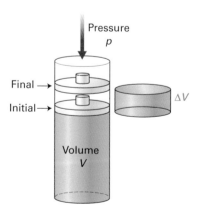

FIGURE 5.2 Finding the work done in expansion.

5

We define **enthalpy H** as

$$H = U + pV \tag{4}$$

Hence $\Delta H = \Delta U + p\Delta V + V\Delta p$ \hfill (5)

Substituting equation (3) into equation (5) we find

$$\Delta H = q - p\Delta V + p\Delta V + V\Delta p = q + V\Delta p \tag{6}$$

This does not seem to be much of an improvement, until we realize that, if the pressure is constant (a reasonable assumption given the ambient pressure in a lab), the second term becomes zero and the final result is that the enthalpy change is given by

$$\Delta H = q \text{ (at constant pressure)} \tag{7}$$

The enthalpy change is equal to the heat added to a system at constant pressure. The word 'enthalpy' comes from the Greek for 'inner warmth'—a very appropriate description.

At university, this derivation will be made rigorous by moving to the infinitesimal limit. Using the language of differentiation, equation (5) becomes

$$dH = dU + pdV + Vdp \tag{8}$$

In the infinitesimal limit, the equation becomes exact.

At university, you will often see this written as $\Delta H = q_p$.

Measuring enthalpy changes

- We can measure enthalpy changes using a **calorimeter**.
- The heat *released* q in a calorimeter is given by $q = mc\Delta T$ where
  - m is the mass of liquid being heated,
  - c is the specific heat capacity of the liquid (4.18 J K^{-1} g^{-1} for water),
  - ΔT is the temperature rise.

See A Deeper Look 5.2 for more about the heat capacity.

A DEEPER LOOK 5.2

What is heat capacity?

The **heat capacity C** of a substance is the heat needed to raise its temperature by 1 degree and is defined by the equation

$$C = q/\Delta T \tag{1}$$

where q is the heat added and ΔT is the temperature rise.

We frequently express the heat capacity in terms of the **specific heat capacity** c, which is the heat capacity per unit mass (measured in J K^{-1} g^{-1}):

$$C = mc \tag{2}$$

Rearranging equation (1) above, and then using equation (2), gives the equation familiar pre-university that

$$q = mc\Delta T \tag{3}$$

At university, you will also frequently come across the **molar heat capacity**, which is the heat capacity per mole (measured in J K^{-1} mol^{-1}). If you are given the molar heat capacity, you need to convert the mass of the substance being heated into its amount in moles (by dividing its mass by its molar mass).

Standard enthalpy change

5

The **standard enthalpy change** $\Delta H^{\ominus}$ is the

- enthalpy change per mole of reaction as written
- when reactants in their standard states form products in their standard states
- at 1 bar (100 kPa)

> You may have encountered the similar sounding expresssion 'enthalpy change when one mole'. This gets the units wrong as it would give a value in kJ rather than kJ mol^{-1}

> Pre-university, you might have come across the expression 'under standard conditions' rather than 'in their standard states'. This fails to highlight the central importance of the concept of the standard state.

> You may have been told that the standard enthalpy change is measured at 298 K. In fact 298 K is simply the *most common* temperature at which the values are measured, when the symbol should be written as $\Delta H^{\ominus}$(298 K).

- The **standard state** of a substance is the pure substance at a pressure of 1 bar. The chapter on spontaneous change (Section 9.1) explains the importance of this fact.

- The **standard enthalpy change of formation** $\Delta_f H^{\ominus}$ is the enthalpy change per mole when a compound is formed **from its *elements*,** all substances being in their standard states at 1 bar.

  - The value for an element is zero, by definition.

  > Although standard enthalpies of formation are available for most substances, a few may not exist at a pressure of 1 bar (such as liquid carbon dioxide).

This symbol, as for the others below, used to be written $\Delta H^{\ominus}_f$.

- The **standard enthalpy change of combustion** $\Delta_c H^{\ominus}$ is the enthalpy change per mole when a substance is **fully combusted in oxygen,** all substances being in their standard states at 1 bar.

5.2 HESS'S LAW AND ENTHALPY CYCLES

- **Hess's law** (found by Germain Hess in 1840) states that the **standard enthalpy change** for a reaction is **independent of the route** taken from reactants to products.

  > The **First Law of thermodynamics** states that energy can neither be created nor destroyed. An alternative definition, using the terminology introduced

earlier (A Deeper Look 5.1), is that **the internal energy of an isolated system is constant**. The latter would be a preferable definition at university.

– We can use Hess's law to find the **standard enthalpy change of reaction $\Delta H^{\ominus}$** (often written $\Delta_r H^{\ominus}$) from the standard enthalpy changes of formation of the products and reactants:

○ $\Delta H^{\ominus} = \Sigma \, \Delta_f H^{\ominus}$ **(products)** $- \Sigma \, \Delta_f H^{\ominus}$ **(reactants)**

■ The summation sign indicates that we need to multiply by the mole ratios for each substance (Section 3.3).

By doing this, you are effectively 'unforming' the reactants, turning them back into their elements, and then forming the elements into products.

So for the reaction of sodium oxide with water to form sodium hydroxide

$$Na_2O(s) + H_2O(l) \rightarrow 2NaOH(aq)$$

$$\Delta H^{\ominus} = 2\Delta_f H^{\ominus}(NaOH,aq) - \Delta_f H^{\ominus}(Na_2O,s) - \Delta_f H^{\ominus}(H_2O,l)$$

• We can use **enthalpy cycles** to find standard enthalpy changes that are hard to measure experimentally.

– One example of an enthalpy cycle is a **Born–Haber cycle**, which describes the formation of an ionic compound from its elements.

Born–Haber cycles

• We can imagine the formation of an ionic compound such as sodium chloride as a series of steps, each of which has a particular standard enthalpy change. Summing up all the standard enthalpy changes must then give the overall standard enthalpy change of formation. Figure 5.3 shows the Born–Haber cycle for sodium chloride.

• The **atomization enthalpy** is the standard enthalpy change accompanying the formation of a gaseous atom from

– either a solid (as for sodium), see step 1 in Figure 5.3:

$$Na(s) \rightarrow Na(g)$$

FIGURE 5.3 The Born–Haber cycle for sodium chloride. The direction of each arrow in this diagram signifies the sign of the enthalpy change: upwards represents an endothermic step; downwards an exothermic step. All figures are in kJ mol^{-1}.

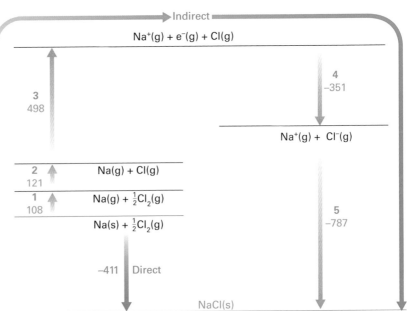

- or a gas containing molecules (as for chlorine), see step 2 in Figure 5.3:

$$\frac{1}{2}Cl_2(g) \rightarrow Cl(g)$$

- Note that it is the value *per mole of atoms*.

- The **first ionization enthalpy** is the standard enthalpy change accompanying the removal of one electron from an atom in the gas phase (Section 1.5); see step 3 in Figure 5.3:

$$E(g) \rightarrow E^+(g) + e^-(g)$$

- The **second ionization enthalpy** is the value for the removal of a second electron.

- The **electron-gain enthalpy** is the standard enthalpy change accompanying the addition of one electron to an atom in the gas phase (see Taking the Next Step 5.2); see step 4 in Figure 5.3:

$$E(g) + e^-(g) \rightarrow E^-(g)$$

An obsolescent name for the electron-gain enthalpy is the **electron affinity**. The reason that the term 'electron affinity' is strongly discouraged is that some books use a positive value for the exothermic process of adding one electron to chlorine. By the acquisitive convention (Section 5.1), the electron-gain enthalpy for an exothermic process *must* be negative.

- The **second electron-gain enthalpy** is the value for the addition of a second electron.

- The values of the electron-gain enthalpy for the halogens are negative (the attachment is exothermic). The value for the second electron-gain enthalpy for oxygen is *positive* because the electron is repelled by the negative charge on O^-: the O^{2-} ion is *not* stable in a vacuum.

TAKING THE NEXT STEP 5.2

Electron-gain enthalpy

For many decades, the most difficult value to measure in the Born–Haber cycle for sodium chloride (Figure 5.3) was the electron-gain enthalpy for chlorine. Most other values in the cycle could be measured with an accuracy of one or two kJ mol^{-1}. The correct value for the electron-gain enthalpy for chlorine was finally measured as -351 kJ mol^{-1}. A necessary consequence of the previous inaccurate value was that the lattice enthalpy for sodium chloride was also in error by the same amount. The commonly used *Chemistry Data Book* (1st edition 1969) by Stark and Wallace has values out by about 15 kJ mol^{-1} for both these values.

- The **lattice formation enthalpy** is the standard enthalpy change accompanying the formation of a solid crystal lattice from a gas of ions; see step 5 in Figure 5.3:

$$Na^+(g) + Cl^-(g) \rightarrow NaCl(s)$$

- The **lattice dissociation enthalpy** is the standard enthalpy change accompanying the breaking of a solid crystal lattice into separate ions in the gas phase:

$$NaCl(s) \rightarrow Na^+(g) + Cl^-(g)$$

  - The lattice dissociation enthalpy is equal in magnitude but opposite in sign to the lattice formation enthalpy.

  - The term **'lattice enthalpy'** should be interpreted as the lattice *dissociation* enthalpy.

- Lattice enthalpies increase with increasing charge on the ions, see Table 5.1.

- Lattice enthalpies decrease with increasing size of the ion; see Table 5.2.

Pre-university, you may have been allowed to use the term 'lattice enthalpy' to mean *either* lattice dissociation *or* lattice formation. However, such mixed usage leads to absurd answers if the wrong sign is used.

5.3 BOND ENTHALPY

- The **bond enthalpy** (or **bond dissociation enthalpy**) is the standard enthalpy change accompanying the breaking of a specific A – B bond into an A atom and a B atom, all in the gas phase:

$$A - B(g) \rightarrow A(g) + B(g)$$

  - Note how bond enthalpy and lattice enthalpy both refer to a *dissociation* process.

- The **mean bond enthalpy** is the average value of the bond enthalpies for a particular bond in a range of molecules containing that bond. Table 5.3 gives some values.

- We can make **rough calculations** of the standard enthalpy change for a reaction by subtracting the sum of the bond enthalpies of the bonds made from the sum of the bond enthalpies of the bonds broken.

The value for one particular bond may be *significantly* different from another. The Cl–F bond enthalpy in the ClF molecule itself is 249 kJ mol^{-1}, whereas the value in the ClF_5 molecule is 142 kJ mol^{-1}.

TABLE 5.1 Lattice enthalpies increase with increasing charge on the ions. MgO is $Mg^{2+}O^{2-}$, whereas NaF is Na^+F^-.

NaF$\Delta_{lat}H^{\ominus}(298\,K) = +926\,kJ\,mol^{-1}$
MgO $\Delta_{lat}H^{\ominus}(298\,K) = +3800\,kJ\,mol^{-1}$

TABLE 5.2 Lattice enthalpies decrease with increasing size of the ion. See Section 13.2 for the relative sizes of the anions.

NaF $\Delta_{lat}H^{\ominus}(298\,K) = +926\,kJ\,mol^{-1}$
NaCl $\Delta_{lat}H^{\ominus}(298\,K) = +787\,kJ\,mol^{-1}$
NaBr $\Delta_{lat}H^{\ominus}(298\,K) = +752\,kJ\,mol^{-1}$
NaI $\Delta_{lat}H^{\ominus}(298\,K) = +705\,kJ\,mol^{-1}$

TABLE 5.3 Average (mean) bond enthalpies.

Bond	Average bond enthalpy/kJ mol^{-1}
C–F	484
C–Cl	338
C–Br	276
C–I	238
C–H	413
C–C	348
C=C	612
C≡C	838
N–H	388
P–H	322
O–H	463
S–H	338
C=O	743
O=O	496
H–H	436

Worked example: What is the standard enthalpy change for the combustion of hydrogen gas to form gaseous water?

Answer: The equation is

$$H_2(g) + \frac{1}{2}O_2(g) \rightarrow H_2O(g)$$

$$\text{Bonds broken} = H–H + \frac{1}{2} \times O=O \text{ taking in } 436 + \frac{1}{2} \times 496 = 684 \text{ kJ mol}^{-1}$$

Bonds made $= 2 \times O–H$ giving out $2 \times 463 = 926 \text{ kJ mol}^{-1}$

Standard enthalpy change $= 684 \text{ kJ mol}^{-1} - 926 \text{ kJ mol}^{-1} = -242 \text{ kJ mol}^{-1}$

- When there is a **major discrepancy** between the measured standard enthalpy change and the value calculated from bond enthalpies, some assumption made about the bonding present is false. See for example the discussion of the delocalization enthalpy of benzene (Section 19.1).

5.4 SOLUTION ENTHALPY AND HYDRATION ENTHALPY

- The **solution enthalpy** is the standard enthalpy change accompanying the dissolution of a solid in a large excess of water.

- The **hydration enthalpy** is the standard enthalpy change accompanying the formation of a hydrated ion from an ion in the gas phase.
- We can find the solution enthalpy by adding together the endothermic lattice enthalpy of the solid and the exothermic hydration enthalpies for the ions formed on its dissociation (taking into account the mole ratios). Sections 12.4 and 13.5 show some example calculations.

CHEMICAL EQUILIBRIUM

6.1 INTRODUCTION TO EQUILIBRIA AND EQUILIBRIUM CONSTANTS

- **Equilibrium** occurs when *the **rate** of the forward reaction **equals** the rate of the backward reaction*.

- Once equilibrium has been reached, *the **concentrations** of all the substances remain **constant***.

- Both forward and backward reactions still occur (chemical equilibrium is **dynamic**); but there is no net change in concentration once equilibrium is reached.

- A **double-headed arrow** ($\rightleftharpoons$) indicates an equilibrium reaction.

Section 9.3 explains the central importance of the Gibbs energy minimum when explaining equilibrium.

Equilibrium constants

- A **homogeneous equilibrium** has all the reactants and products in the same phase.

- We define the **equilibrium constant** in terms of **concentration**, K_c, for the equilibrium

  $a\text{A} + b\text{B} \rightleftharpoons c\text{C} + d\text{D}$ as follows:

  $$K_c = \frac{[\text{C}]^c[\text{D}]^d}{[\text{A}]^a[\text{B}]^b}$$

  - where the concentrations in this equation are those *when equilibrium is reached*.
  - The units of K_c will depend on the values of *a*, *b*, *c*, and *d*. See the worked examples below (Section 6.2) and Taking the Next Step 6.1.

  Example:

  - For the **esterification** reaction forming ethyl ethanoate (Section 23.4):

    $CH_3COOH + CH_3CH_2OH \rightleftharpoons CH_3COOCH_2CH_3 + H_2O$

    $$K_c = \frac{[CH_3COOCH_2CH_3][H_2O]}{[CH_3COOH][CH_3CH_2OH]}$$

'Phase' is a better word than 'state', as two immiscible liquids form a two-phase system despite both being in the liquid state.

TABLE 6.1 Data for five experiments (at 373 K) showing initial concentrations of acid and alcohol, equilibrium concentrations of ester (and water), and associated K_c values.

Experiment	$[Acid]_0/$ mol dm^{-3}	$[Alcohol]_0/$ mol dm^{-3}	$[Ester]_{eq}/$ mol dm^{-3}	K_c
1	1.00	0.18	0.171	3.9
2	1.00	0.50	0.420	3.8
3	1.00	1.00	0.667	4.0
4	1.00	2.00	0.858	4.5
5	1.00	8.00	0.966	3.9

- K_c is independent of concentration. Table 6.1 shows that, whatever the initial concentrations, the esterification reaction reaches an equilibrium composition where the concentrations are related by K_c.
- K_c *does* depend on temperature.
- We define the **equilibrium constant** in terms of **pressure, K_p,** for the gaseous equilibrium

  $aA(g) + bB(g) \rightleftharpoons cC(g) + dD(g)$ as follows:

  $$K_p = \frac{p(C)^c \, p(D)^d}{p(A)^a \, p(B)^b}$$

  - where the partial pressures in this equation are those *when equilibrium is reached*.
  - The units of K_p will depend on the values of a, b, c, and d. See the worked examples below (Section 6.2) and Taking the Next Step 6.1.
  - In the same way that K_c is independent of concentration, K_p is independent of pressure. Whatever the initial pressures, a reaction reaches an equilibrium composition where the pressures are related by K_p.
  - K_p *does* depend on temperature, as is the case for K_c.
- The **partial pressure** of gas A, $p(A)$, in a mixture of gases is the pressure it would exert if it alone occupied the total volume occupied by the gas mixture; the partial pressure of A is equal to its mole fraction $x(A)$ multiplied by the total pressure $p(total)$:
  - $p(A) = x(A) \, p(total)$
- The **mole fraction**, as its name implies, is the fraction of the total amount in moles made up of one particular gas:
  - $x(A) = \dfrac{n(A)}{n(total)}$
    - The sum of all the mole fractions is therefore always equal to 1.

Example:

For the ammonia synthesis (Section 6.3) $N_2(g) + 3H_2(g) \rightleftharpoons 2NH_3(g)$

$$K_p = \frac{p(NH_3)^2}{p(N_2)p(H_2)^3}$$

Taking the Next Step 6.1 explains how units are treated differently at university for both K_c and K_p.

TAKING THE NEXT STEP 6.1

A question of units

Because the standard Gibbs energy change is proportional to the natural logarithm of the equilibrium constant (Section 9.3), scientists have devised the **thermodynamic equilibrium constant K**, which always has no units. (It is impossible to take the natural logarithm of a quantity with units.) This will be used consistently at university.

We define K_c above in terms of the concentrations [X] and, as such, express each concentration in mol dm^{-3}. The definition of the thermodynamic equilibrium constant K is exactly the same, except that we use $[X]/c^{\ominus}$ (where $c^{\ominus}$ is the standard concentration of **1 mol dm^{-3}**) in place of [X].

Similarly, for K_p, we replace the partial pressure $p(X)$ by $p(X)/p^{\ominus}$ (where $p^{\ominus}$ is the standard pressure of **1 bar**). The definition of the thermodynamic equilibrium constant K is exactly the same otherwise. So for the ammonia synthesis:

$$K = \frac{(p(NH_3)/p^{\ominus})^2}{(p(N_2)/p^{\ominus})(p(H_2)/p^{\ominus})^3} = \frac{p(NH_3)^2 \times (p^{\ominus})^4}{p(N_2)p(H_2)^3 \times (p^{\ominus})^2} = \frac{p(NH_3)^2 \times (p^{\ominus})^2}{p(N_2)p(H_2)^3}$$

A word of caution here: although concentrations are almost always given in mol dm^{-3}, hence dividing by 1 mol dm^{-3} does not change any numbers, pressures are *not* always given in bar, and we may need to make a conversion to find the correct equilibrium constant in the end. Therefore, using the data in worked example 2 on ammonia immediately below, we need to use the fact that 1 bar = 0.987 atm to find

$$K = \frac{p(NH_3)^2 \times (p^{\ominus})^2}{p(N_2)p(H_2)^3} = \frac{35.2^2 \times 0.987^2}{41.2 \times 123.6^3} = 1.55 \times 10^{-5}$$

6.2 EQUILIBRIUM CALCULATIONS

Most pre-university calculations revolve around defining the equilibrium constant for a particular reaction and then substituting the values given for the concentrations or pressures.

Worked examples on equilibrium calculations

1 The equation for the gaseous reaction between hydrogen and iodine is

$$H_2(g) + I_2(g) \rightleftharpoons 2HI(g)$$

At equilibrium at 490 °C, the concentrations are 2.18 mol dm^{-3} H_2, 2.87 mol dm^{-3} I_2, and 16.9 mol dm^{-3} HI. Find K_c.

Solution:

$$K_c = \frac{[HI]^2}{[H_2][I_2]}$$

$$= \frac{(16.9\,\text{mol dm}^{-3})^2}{(2.18\,\text{mol dm}^{-3})(2.87\,\text{mol dm}^{-3})}$$

$$= \frac{(16.9)^2}{2.18 \times 2.87}$$

$$= 46$$

2 In the Haber–Bosch process for manufacturing ammonia (Section 6.3), at a temperature of 500 °C and a pressure of 200 atm, the partial pressures are: N_2 41.2 atm, H_2 123.6 atm, NH_3 35.2 atm. (Figure 6.2 shows how the percentage of ammonia changes with pressure at various temperatures.) Find K_p.

Solution:

$$K_p = \frac{p(NH_3)^2}{p(N_2)p(H_2)^3}$$

$$= \frac{(35.2\,\text{atm})^2}{(41.2\,\text{atm})(123.6\,\text{atm})^3}$$

$$= 1.6 \times 10^{-5}\,\text{atm}^{-2}$$

3 Let us think about the third line of data in Table 6.1. Because the reaction started with only acid and alcohol present, the same amount of water must be formed as ester (0.667 mol). To form one molecule of ester and one of water, one molecule of acid and one of alcohol must react. Therefore, the amount of acid and the amount of alcohol left at equilibrium is (1.00 − 0.667) = 0.333 mol.

$$K_c = \frac{\left[CH_3COOCH_2CH_3\right][H_2O]}{\left[CH_3COOH\right][CH_3CH_2OH]}$$

$$K_c = \frac{(0.667\,\text{mol}/V)(0.667\,\text{mol}/V)}{(0.333\,\text{mol}/V)(0.333\,\text{mol}/V)}$$

$$= \frac{(0.667)^2}{(0.333)^2}$$

$$= 4.0$$

Notice how the units cancel in both examples 1 and 3 (and the volume V cancels in 3) because two molecules are being formed from two molecules.

A Deeper Look 6.1 explains how to do the harder calculations discussed at university.

A DEEPER LOOK 6.1

How can we make an equilibrium table?

When questions become more difficult, such as attempting to find the composition of an equilibrium mixture, we need to set up an **equilibrium**

TABLE 6.2 The equilibrium table for the ammonia synthesis.

Species	N_2	H_2	NH_3
p(initial)/bar	1	3	0
change/bar	$-x$	$-3x$	$+2x$
p(equil)/bar	$1-x$	$3-3x$	$2x$

table, which shows how the initial composition changes as the reaction approaches equilibrium.

This is also called an **ICE table** (Initial, Change, Equilibrium).

A typical question is as follows. In the ammonia synthesis (Section 6.3), 1 bar nitrogen reacts with 3 bar hydrogen to form ammonia, in a reactor of fixed volume. If x moles of nitrogen react, $3x$ moles of hydrogen must react as well (given the 1:3 ratio in the chemical equation) to form $2x$ moles of ammonia. The equilibrium constant K_p at the temperature of the experiment is 976 bar^{-2}. So we can construct the equilibrium table (Table 6.2) as shown above.

We can then write the equilibrium constant as

$$K_p = \frac{p(NH_3)^2}{p(N_2)p(H_2)^3}$$

$$= \frac{(2x)^2 \text{ bar}^{-2}}{(1-x)(3-3x)^3}$$

$$= \frac{4x^2 \text{ bar}^{-2}}{27(1-x)^4}$$

$$\text{So } 976 = \frac{4x^2}{27(1-x)^4}$$

In this case, the equation solves to give $x = 0.895$. So $p(NH_3) = 1.79$ bar, $p(N_2) = 0.105$ bar, and $p(H_2) = 0.315$ bar. It is important to check that the answers make sense by calculating a value for K_p: here 976 bar^{-2}.

The example above involves pressures, but concentrations can also be used.

6.3 LE CHATELIER'S PRINCIPLE

- **Le Chatelier's principle** states that if a system at equilibrium is subjected to a **small change in conditions**, the equilibrium tends to *shift* to *minimize* the effect of the change.

Le Chatelier's principle applied to changing concentration and pressure

- **Increasing the concentration of one of the reactants** causes the equilibrium to **shift towards the products**; increasing the concentration of one of the products causes the equilibrium to shift towards the reactants.
  - We can also calculate the magnitude of the shift by using the fact that the equilibrium constant K_c does not change when the concentrations change.

* **Increasing the pressure** on a gaseous equilibrium causes the equilibrium to **shift towards the side with fewer moles of gas**.
  - Taking the Next Step 6.2 proves this using the fact that the equilibrium constant K_p does not change when the pressure changes.

TAKING THE NEXT STEP 6.2

Calculating the effect of a change in pressure quantitatively

We can rewrite the expression for K_p for the ammonia synthesis using the equation linking the partial pressures and the mole fractions (Section 6.1):

$$K_p = \frac{p(NH_3)^2}{p(N_2)p(H_2)^3}$$

$$= \frac{x(NH_3)^2\, p(\text{total})^2}{x(N_2)p(\text{total})x(H_2)^3\, p(\text{total})^3}$$

Dividing top and bottom by $p(\text{total})^2$:

$$K_p = \frac{x(NH_3)^2}{x(N_2)x(H_2)^3\, p(\text{total})^2}$$

$$= \frac{K_x}{p(\text{total})^2}$$

$$\text{where } K_x = \frac{x(NH_3)^2}{x(N_2)x(H_2)^3}$$

The equilibrium constant K_p *does not change as pressure changes*. So if $p(\text{total})^2$ increases, K_x must increase to the same extent. However, the sum of the mole fractions is always 1 (Section 6.1). Therefore the mole fraction of ammonia, $x(NH_3)$, must increase, while the mole fractions of nitrogen, $x(N_2)$, and hydrogen, $x(H_2)$, decrease. This mirrors the conclusion suggested by Le Chatelier's principle, but the equation above allows us to calculate the change quantitatively.

Le Chatelier's principle applied to changing temperature

* The explanation for the next two bullet points follows from applying Le Chatelier's principle:
  - **Increasing the temperature** shifts an equilibrium in the **endothermic** direction.
  - **Decreasing the temperature** shifts an equilibrium in the **exothermic** direction.

Figure 6.1 illustrates this principle applied to the equilibrium between NO_2 and its dimer N_2O_4:

$$2NO_2(g) \rightleftharpoons N_2O_4(g)$$

The reaction is exothermic in the forward direction, because a bond is formed between the two nitrogen atoms and no other bonds are broken. So

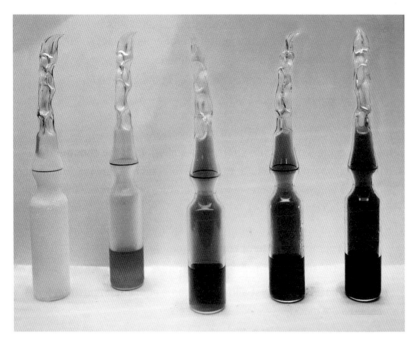

FIGURE 6.1 The dimerization of nitrogen dioxide at various temperatures. When hot (right) red-brown NO_2 predominates; when colder (middle) there is much less NO_2. When very cold (left) this equilibrium system appears almost colourless (nearly 100% N_2O_4).

increasing the temperature shifts the equilibrium towards the reactants: more red-brown NO_2 forms at higher temperature.

- Note carefully that the **equilibrium constant will be different** at different temperatures.

A Deeper Look 9.2 shows how to find the effect of changing temperature on the equilibrium constant quantitatively.

The effect of a catalyst

- **A catalyst *does not affect the equilibrium position*.**
- **A catalyst allows the same equilibrium position to be reached *more quickly*.**

Applying Le Chatelier's principle to the Haber–Bosch ammonia synthesis

- The **Haber–Bosch process** (also called the **Haber process**) is the name given to the industrial synthesis of ammonia:
  - $N_2(g) + 3H_2(g) \rightleftharpoons 2NH_3(g)$
- Increasing the pressure causes the equilibrium to shift towards the side with fewer moles of gas (here, the products). Figure 6.2 shows that **increasing pressure will increase the yield** in the ammonia synthesis.

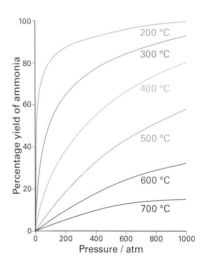

FIGURE 6.2 The quantitative effect of pressure on ammonia synthesis. At equilibrium, a larger percentage of ammonia is present at high pressure. In practice, there is a trade-off between initial costs and yield.

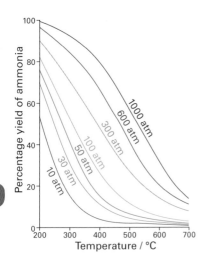

FIGURE 6.3 A graph of percentage yield against temperature at various pressures for the production of ammonia from nitrogen and hydrogen.

Taking the Next Step 21.1 describes the application of Le Chatelier's principle to the manufacture of ethanol.

- However, the costs of maintaining the high pressure (energy costs, use of special steels, and preparation for emergencies) escalate. Hence a **compromise pressure** is required.

  ○ In the UK, the compromise pressure is typically **250 atm**.

  > The compromise pressure used in France is typically 1000 atm. The costs of construction and maintenance of safety measures increase but the long-term profit will be higher as more ammonia is made.

- Increasing the temperature shifts the equilibrium in the **endothermic** direction (towards the reactants, as the forward reaction is exothermic). Figure 6.3 shows that **increasing temperature will decrease the yield** in the ammonia synthesis.

  - However, at lower temperatures the reaction will occur *more slowly*. Hence a **compromise temperature** is required.

    ○ In the UK, the compromise temperature is typically 450 °C.

- The presence of a catalyst allows the same equilibrium position to be reached *more quickly*.

  - The catalyst used is **iron**.

    > A **promoter** increases the rate of reaction even more, although it does not catalyse the reaction itself. Promoters improve both the surface area and the electronic structure of the catalyst. The promoters in the Haber–Bosch ammonia synthesis are the oxides of potassium, calcium, and aluminium.

ACID–BASE EQUILIBRIUM

7.1 INTRODUCTION TO ACID–BASE EQUILIBRIA

- **Acid–base equilibrium** involves **transfer of protons** according to the **Brønsted–Lowry theory** (but see also Section 7.6 at the end of this chapter).

- A **Brønsted acid** is a **proton donor**, as for ethanoic acid in the following equilibrium:
  - $CH_3COOH + H_2O \rightleftharpoons H_3O^+ + CH_3COO^-$
  - CH_3COOH and CH_3COO^- form a **conjugate acid–base pair** as they are related by the transfer of a proton (as are H_3O^+ and H_2O).

- A **Brønsted base** is a **proton acceptor**, as for ammonia in the following equilibrium:
  - $NH_3 + H_2O \rightleftharpoons NH_4^+ + OH^-$
  - NH_4^+ and NH_3 form a conjugate acid–base pair (as do H_2O and OH^-).
  - An **alkali** is a water-soluble base.

To *simplify* the connection between the parent acid and its ion, we have ignored the delocalization in the ion, as is frequently done in writing such equilibria. The anion's formula should be written $CH_3CO_2^-$. (See Section 2.3.)

7.2 pH CALCULATIONS FOR STRONG ACIDS AND STRONG BASES

- **pH $= -\log_{10}[H^+]$** with $[H^+]$, the hydrogen ion concentration, measured in mol dm^{-3}.
  - The hydrogen ion concentration in mol dm^{-3} is given by $[H^+] = 10^{-pH}$.
  - We can **measure** pH using indicators (see Section 7.4), a pH meter, or even universal indicator/pH paper.

- A **strong acid** is an acid that is **fully ionized** in aqueous solution: the equilibrium reaction for the ionization lies fully to the right. Consequently, the hydrogen ion concentration equals the given concentration of the acid. Strong acids include hydrochloric acid and nitric acid.

The Dane Søren Sørensen introduced the concept of pH in 1909 to enable brewers to measure the acidity of their lager and hence provide the optimum conditions for fermentation.

- For a strong acid of concentration 1 mol dm⁻³, pH = −log₁₀(1) = 0.0.
- For a strong acid of concentration 0.1 mol dm⁻³, pH = −log₁₀(0.1) = 1.0.
- For a strong acid of concentration 0.01 mol dm⁻³,
  pH = −log₁₀(0.01) = 2.0.
- Each dilution by a factor of ten increases the pH by one unit.
  - The number of *decimal places* for pH should equal the number of *significant figures* in the concentration (see Taking the Next Step 7.1).

- A **strong base** is a base that is **fully ionized** in aqueous solution. Consequently, the hydroxide ion concentration equals the given concentration of the base. Strong bases include sodium hydroxide and potassium hydroxide.
  - To relate the hydroxide ion and hydrogen ion concentrations, we need to use the ionic product of water.

- We define the **ionic product of water K_w** by
  $K_w = [H^+][OH^-] = 1 \times 10^{-14}$ mol² dm⁻⁶ at 298 K.
  - For a strong base of concentration 1 mol dm⁻³, $[OH^-] = 1$ mol dm⁻³, $[H^+] = 1 \times 10^{-14}$ mol dm⁻³, and pH = −log₁₀(1×10⁻¹⁴) = 14.0.
  - For a strong base of concentration 0.1 mol dm⁻³, $[OH^-] = 0.1$ mol dm⁻³, $[H^+] = 1 \times 10^{-13}$ mol dm⁻³, and pH = −log₁₀(1×10⁻¹³) = 13.0.
  - For a strong base of concentration 0.01 mol dm⁻³, $[OH^-] = 0.01$ mol dm⁻³, $[H^+] = 1 \times 10^{-12}$ mol dm⁻³, and pH = −log₁₀(1×10⁻¹²) = 12.0.
  - Each dilution by a factor of ten decreases the pH by one unit.
  - Hence **the typical pH range in aqueous solutions is from 0 to 14**.
  - **pK_w = −log₁₀K_w** with K_w measured in mol² dm⁻⁶ (pre-university).
  - **pK_w = 14.0 at 298 K**.

TAKING THE NEXT STEP 7.1

Decimal places for pH

The number immediately in front of the decimal point in a pH value simply gives the power of ten: all hydrogen ion concentrations between 1 mol dm⁻³ and 0.1 mol dm⁻³ will give a pH value starting with a zero. For example, if $[H^+] = 0.5$ mol dm⁻³ then pH = 0.3 while if $[H^+] = 0.2$ mol dm⁻³ then pH = 0.7.

$[H^+] = 0.50$ mol dm⁻³ gives pH = 0.30, if $[H^+] = 0.51$ mol dm⁻³ then pH = 0.29, and if $[H^+] = 0.55$ mol dm⁻³ then pH = 0.26. Notice how the second decimal place changes with the changing second significant figure in the concentration.

Pre-university, you will often have been asked for two decimal places *whatever* the number of significant figures given for the concentration. As usual, at university you need to think carefully about the number of decimal places you choose to use.

7.3 pH CALCULATIONS FOR WEAK ACIDS

- A **weak acid**, such as ethanoic acid, is an acid that is **only partially ionized** in aqueous solution:
  - $CH_3COOH + H_2O \rightleftharpoons H_3O^+ + CH_3COO^-$

- Similarly a **weak base**, such as aqueous ammonia, is a base that is **only partially ionized** in aqueous solution:
  - $NH_3 + H_2O \rightleftharpoons NH_4^+ + OH^-$

- We can calculate the extent of the ionization using the **acid ionization constant (acid dissociation constant)** K_a:
  - $HA \rightleftharpoons H^+ + A^-$

  $$K_a = \frac{[H^+][A^-]}{[HA]}$$

  - where [HA] is the acid concentration *when equilibrium has been reached*. We measure K_a in mol dm^{-3}; however, see Taking the Next Step 7.2.
  - The smaller K_a is, the weaker the acid is.

> Sulfuric acid H_2SO_4 is a strong acid in its first ionization (forming HSO_4^-) but is a weak acid in its second (forming SO_4^{2-}).

> It is incorrect to say, as is common pre-university, that ammonia partially *dissociates*: instead, it is *gaining* a proton as the ion forms. (It is, however, acceptable to say that a weak *acid* partially dissociates.)

> Notice how we have simplified the formula for the aqueous hydrogen ion from H_3O^+ to H^+, as is commonly done during calculations.

TAKING THE NEXT STEP 7.2

A simple application of equilibrium principles

Applying the rules described in Section 6.1 for writing the equilibrium constant K_c for the

$HA \rightleftharpoons H^+ + A^-$

equilibrium would result in the same equation as shown above. We change the subscript to show that this specifically applies to an equilibrium involving an acid.

At university, K_a will have no units as it is treated as a thermodynamic equilibrium constant (Section 6.1).

- The **A$^-$ ion concentration equals the H$^+$ ion concentration** as both ions are formed in equal quantities when the acid ionizes. Then, assuming that the **acid ionizes very little**, we can equate [HA] at equilibrium to that added at the start. We then find that $K_a = [H^+]^2/[HA]$ so $[H^+] = \sqrt{(K_a[HA])}$

- The quantity pK_a is related to K_a in the same way as pH is related to the hydrogen ion concentration, namely
  - $pK_a = -\log_{10}K_a$ with K_a measured in mol dm^{-3} (pre-university).
  - K_a in mol dm^{-3} is given by $K_a = 10^{-pK_a}$
  - If K_a is 1.74×10^{-5} (as is the case for ethanoic acid), pK_a is 4.76.
    - The higher the pK_a, the weaker the acid is.

> Remember that [H$^+$], [HA], and K_a (pre-university) are all measured in mol dm^{-3}. In order to simplify the rest of this section, the units for the concentrations (and for K_a) are left out.

- So we can finally find the **pH of a weak acid** as follows. For ethanoic acid,

$$K_a = 1.74 \times 10^{-5} \text{ (as shown above)}$$

  - at a concentration of 1 mol dm^{-3}, [HA] = 1,

$$[H^+] = \sqrt{(K_a[HA])}$$
$$= \sqrt{(1.74 \times 10^{-5} \times 1)}$$
$$= 0.0042$$
$$\text{so pH} = -\log_{10}(0.0042)$$
$$= 2.4 \,(1\,dp)$$

  - at a concentration of 0.1 mol dm^{-3}, [HA] = 0.1,

$$[H^+] = \sqrt{(K_a[HA])}$$
$$= \sqrt{(1.74 \times 10^{-5} \times 0.1)}$$
$$= 0.0013$$
$$\text{so pH} = -\log_{10}(0.0013)$$
$$= 2.9 \,(1\,dp)$$

As shown above, a strong acid with these concentrations would produce a larger change, from a pH of 0.0 to a pH of 1.0.

7.4 TITRATIONS

- We use **acid–base titrations** (Section 3.7) to measure the unknown concentration of one solution by reaction with another solution whose concentration we do know (a **standard solution**). We find the volume of the standard solution that exactly neutralizes a measured volume of the unknown solution using an indicator.

- **Indicators** are typically water-soluble weak organic acids that have different colours at different pH values, as shown in Figure 7.1.
  - We call the pH at which the colour changes significantly the indicator's **pK_{in}**. (The pK_{in} is the pK_a of the weak acid used as the indicator.)
  - One pH unit below pK_{in} the indicator will show its 'acidic' colour; one pH unit above pK_{in} the indicator will show its 'basic' colour.

- The pH changes during a titration vary very significantly depending on whether the acid is strong or weak and whether the base is strong or weak.
  - The **equivalence point (or stoichiometric point)** occurs when the acid exactly neutralizes the base.
  - The **end point** occurs when the indicator changes colour. It is the aim of a titration that the end point is as close as possible to the equivalence point. **A weak acid should not be titrated against a weak base**, as no indicator can find the equivalence point.
  - Usually we add the base into a conical flask using a **pipette**, and then run the acid in from a **burette** (Figure 3.7).

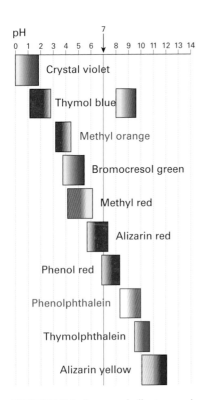

FIGURE 7.1 Common indicators and their ranges.

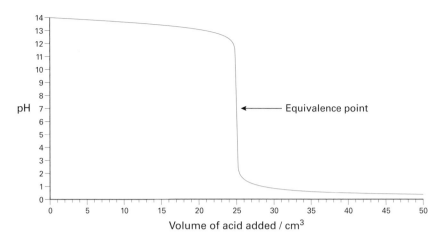

FIGURE 7.2 The equivalence point for a strong acid/strong base titration (here both at a concentration of 1 mol dm^{-3}) is at pH 7. The pH changes from 11 to 3 very rapidly so phenolphthalein (range pH 8-10), litmus (range pH 6-8), and methyl orange (range pH 3-5) would all be suitable indicators.

- For a **strong acid** reacting with a **strong base**, the pH changes dramatically from about 11 just before neutralization to about 3 just after neutralization, as shown in Figure 7.2. The solution at the equivalence point will have a pH very close to 7.

  > One drop (0.05 cm^3) before neutralization, the total volume will be very close to 50 cm^3, so the strong base is diluted by a factor of 50/0.05 = 1000. Hence the pH falls by three units, from 14 (assuming an initial concentration of 1 mol dm^{-3}) to 11. One drop (0.05 cm^3) after neutralization, the strong acid is diluted by a factor of 1000. Assuming an initial acid concentration of 1 mol dm^{-3}, the pH will be 3.

  - Any common indicator will work satisfactorily for a strong acid/ strong base titration.

- For a **strong acid** reacting with a **weak base**, the pH changes dramatically from a little above 7 just before neutralization to about 3 just after neutralization, as shown in Figure 7.3. The solution at the equivalence point will have a pH around 5.

  - The indicator for a **strong acid/weak base** titration must have a pK_{in} near 5. Figure 7.1 shows that **methyl orange** or bromocresol green would be suitable.

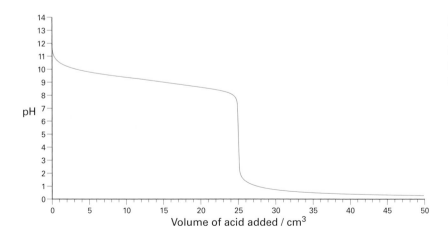

FIGURE 7.3 The titration curve of a strong acid (e.g. hydrochloric acid HCl) with a weak base (here aqueous ammonia NH$_3$). Here both are at a concentration of 1 mol dm^{-3}.

FIGURE 7.4 The titration curve of a weak acid (here ethanoic acid CH_3COOH) with a strong base (e.g. sodium hydroxide NaOH). Here both are at a concentration of 1 mol dm^{-3}.

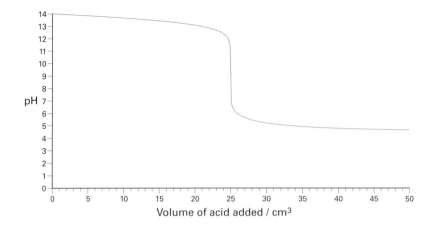

- For a **weak acid** reacting with a **strong base**, the pH changes dramatically from about 11 just before neutralization to a little below 7 just after neutralization, as shown in Figure 7.4. The solution at the equivalence point will have a pH about 9.

  - The indicator for a **weak acid/strong base** titration must have a pK_{in} near 9. Figure 7.1 shows that **phenolphthalein** would be suitable.

    We can work out the pH at 50 cm^3 relatively easily. The first 25 cm^3 of acid neutralizes the base to form salt. Adding a further 25 cm^3 means that the acid and its salt are in exactly equal concentration. The pH of this buffer solution is equal to the pK_a of the acid (4.8 in the case of ethanoic acid): see Section 7.5 on buffer solutions.

7.5 BUFFER SOLUTIONS

- Figure 7.5 shows what would happen if we did the weak acid/strong base titration by adding the *acid* by pipette and running the base in from a burette.

- Figure 7.5 shows how, at the very start of the titration, the pH rises steeply but then the slope levels out significantly as the **buffer zone** is reached. In the buffer zone, the pH changes relatively slowly. Close to neutralization, the slope becomes steeper again, before becoming very steep as neutralization occurs.

- A **buffer solution** resists change in pH when *small* amounts of acid or base are added.

- An **acidic buffer solution** consists of a weak acid and its salt, e.g. CH_3COOH and CH_3COONa (produced during the titration by reaction of the acid with NaOH).

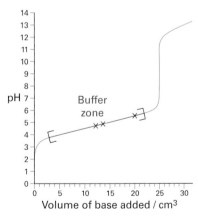

FIGURE 7.5 The titration curve for ethanoic acid against sodium hydroxide (where both have a concentration of 1 mol dm^{-3}). Taking the Next Step 7.3 shows how to calculate the pH values marked in the buffer zone.

- A typical acidic buffer solution contains significant amounts of both CH_3COOH and CH_3COO^- ions.

  - If strong acid is added, the just added H^+ ions react with CH_3COO^- ions to form CH_3COOH, so overall leaving the pH nearly unchanged.

  - If strong base is added, the just added OH^- ions react with CH_3COOH to form CH_3COO^- ions, so overall leaving the pH nearly unchanged.

Taking the Next Step 7.3 shows how to do acidic buffer solution calculations.

Acidic buffer solution calculations

The strategy for doing **buffer solution calculations** involving a certain volume of a weak acid reacted with a certain volume of sodium hydroxide, a strong base, is as follows:

If you find buffer solution calculations difficult, you are not alone! Almost everyone does.

- Calculate the amount in moles of the weak acid at the start, $n(\text{acid})$.
- Calculate the amount in moles of the strong base at the start, $n(\text{NaOH})$.
  - As this is a buffer solution, $n(\text{NaOH})$ is necessarily *smaller* than $n(\text{acid})$.
- **All** the base is neutralized to form the salt NaA, which is fully ionized (Na^+A^-)
  - $n(\text{A}^-) = n(\text{NaOH})$
- **Some** of the acid has been used up in making the salt, so we can find the amount of the weak acid HA remaining by
  - $n(\text{HA}) = n(\text{acid}) - n(\text{NaOH})$
- Going back to the definition of K_a (Section 7.3) and making the hydrogen ion concentration the subject, $[\text{H}^+] = K_a[\text{HA}]/[\text{A}^-]$
  - The ratio of concentrations equals the ratio of the amounts in moles as the solution volume is the same.
- Hence $[\textbf{H}^+] = \boldsymbol{K_a}n(\textbf{HA})/n(\textbf{A}^-)$, from which we can calculate the pH.
  - The most important consequence of this equation is that when there is an *equal* amount of acid and salt (**exactly half-way to neutralization**),
  - $n(\text{HA}) = n(\text{A}^-)$ and $\textbf{pH} = \textbf{p}\boldsymbol{K_a}$

Examples of buffer solution calculations

1 At 25 °C, the $\text{p}K_a$ value for ethanoic acid is 4.76. Calculate the pH of the solution formed when 12.5 cm^3 of 1.00 mol dm^{-3} sodium hydroxide reacts with 25.0 cm^3 of 1.00 mol dm^{-3} ethanoic acid at 25 °C.

$$n(\text{acid}) = \left(\frac{25.0}{1000}\right) \text{dm}^3 \times (1.00\,\text{mol dm}^{-3}) = 2.50 \times 10^{-2}\,\text{mol}$$

$$n(\text{NaOH}) = \left(\frac{12.5}{1000}\right) \text{dm}^3 \times (1.00\,\text{mol dm}^{-3}) = 1.25 \times 10^{-2}\,\text{mol}$$

$$n(\text{A}^-) = n(\text{NaOH}) = 1.25 \times 10^{-2}\,\text{mol}$$
$$n(\text{HA}) = n(\text{acid}) - n(\text{NaOH})$$
$$= 2.50 \times 10^{-2}\,\text{mol} - 1.25 \times 10^{-2}\,\text{mol}$$
$$= 1.25 \times 10^{-2}\,\text{mol}$$

Hence $n(\text{HA}) = n(\text{A}^-)$ and pH = $\text{p}K_a$ = 4.76

Comment: This is exactly half-way to neutralization, so $pH = pK_a$

2 Repeat the above calculation after a further 1.0 cm³ of sodium hydroxide has been added. See Section 7.3 for the way of finding K_a from pK_a.

$$n(\text{acid}) = \left(\frac{25.0}{1000}\right) dm^3 \times (1.00\,mol\,dm^{-3}) = 2.50 \times 10^{-2}\,mol$$

$$n(\text{NaOH}) = \left(\frac{13.5}{1000}\right) dm^3 \times (1.00\,mol\,dm^{-3}) = 1.35 \times 10^{-2}\,mol$$

$$n(A^-) = n(\text{NaOH}) = 1.35 \times 10^{-2}\,mol$$
$$n(\text{HA}) = n(\text{acid}) - n(\text{NaOH})$$
$$= 2.50 \times 10^{-2}\,mol - 1.35 \times 10^{-2}\,mol$$
$$= 1.15 \times 10^{-2}\,mol$$
$$[H^+] = K_a n(\text{HA})/n(A^-)$$
$$= (10^{-4.76}\,mol\,dm^{-3}) \times (1.15 \times 10^{-2}\,mol)/(1.35 \times 10^{-2}\,mol)$$
$$= 1.48 \times 10^{-5}\,mol\,dm^{-3}$$
$$pH = -\log_{10}(1.48 \times 10^{-5}) = 4.83$$

Comment: The pH increases very little when 1 cm³ of strong base is added to 37.5 cm³ of this buffer solution. Had 1 cm³ of sodium hydroxide been added to 37.5 cm³ of water, the initial pH would have been 7.0. The base would be diluted by a factor of 38.5, so the hydroxide ion concentration would be $2.60 \times 10^{-2}\,mol\,dm^{-3}$ and the pH would rise to 12.4, a *dramatically* greater change of 5.4 rather than $4.83 - 4.76 = 0.07$. The buffer solution has resisted pH change very effectively.

3 Repeat the above calculation when a total of 20.0 cm³ of sodium hydroxide has been added.

$$n(\text{acid}) = \left(\frac{25.0}{1000}\right) dm^3 \times (1.00\,mol\,dm^{-3}) = 2.50 \times 10^{-2}\,mol$$

$$n(\text{NaOH}) = \left(\frac{20.0}{1000}\right) dm^3 \times (1.00\,mol\,dm^{-3}) = 2.00 \times 10^{-2}\,mol$$

$$n(A^-) = n(\text{NaOH}) = 2.00 \times 10^{-2}\,mol$$
$$n(\text{HA}) = n(\text{acid}) - n(\text{NaOH})$$
$$= 2.50 \times 10^{-2}\,mol - 2.00 \times 10^{-2}\,mol$$
$$= 5.00 \times 10^{-3}\,mol$$
$$[H^+] = K_a n(\text{HA})/n(A^-)$$
$$= (10^{-4.76}\,mol\,dm^{-3}) \times (5.00 \times 10^{-3}\,mol)/(2.00 \times 10^{-2}\,mol)$$
$$= 4.34 \times 10^{-6}\,mol\,dm^{-3}$$
$$pH = -\log_{10}(4.34 \times 10^{-6}) = 5.36$$

Comment: See the points picked out in Figure 7.5 at 12.5, 13.5, and 20.0 cm³.

- A **basic buffer solution** consists of a weak base and its salt, e.g. NH_3 and NH_4Cl. Figure 7.6 shows a basic buffer zone.

7.6 LEWIS ACIDS AND BASES

Taking the Next Step 7.4 introduces another very important concept: Lewis acids and bases.

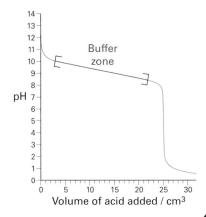

FIGURE 7.6 The titration curve for aqueous ammonia against hydrochloric acid (where both have a concentration of 1 mol dm^{-3}). The pH at the centre of the buffer zone is equal to the pK_a of the ammonium ion NH_4^+ which is 9.3 (see Taking the Next Step 7.3).

TAKING THE NEXT STEP 7.4

Lewis acids and bases

An archetypal reaction between an acid and a base is

$$H^+ + OH^- \rightarrow H_2O$$

Brønsted–Lowry theory (Section 7.1) regards this as the transfer of a proton to the hydroxide ion. An alternative, equally valid view would be that the hydroxide ion transfers a pair of electrons to the proton to make the new O–H bond.

This leads to an alternative theory of acid–base reactions introduced by Gilbert Lewis:

A **Lewis acid** is an **electron-pair acceptor**.

A **Lewis base** is an **electron-pair donor**.

All Brønsted acids are Lewis acids. However, we can extend the concept of Lewis acids to reactions *that do not involve protons*. So we can understand that complexation reactions (reactions that form complexes or complex ions, Section 15.1) are Lewis acid–base reactions. The central copper(II) ion acts as the Lewis acid and the water molecules act as Lewis bases in the complex shown in Figure 7.7. See also Sections 15.2 and 15.3.

Boron trifluoride reacts with ammonia (Section 2.5), as illustrated in Figure 7.8, in a very similar way to the reaction between hydrogen chloride and ammonia.

Furthermore, we can classify the various different catalysts used in Friedel–Crafts acylation (Section 19.4) as Lewis acids.

The concept of Lewis acidity/basicity is therefore especially helpful at university to interrelate similar types of reactions in both inorganic and organic chemistry.

The definitions of Lewis acids/bases and electrophiles/nucleophiles (Section 16.4) are closely similar.

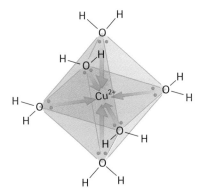

FIGURE 7.7 The hexaaquacopper(II) complex ion.

FIGURE 7.8 (a) Boron trifluoride reacts with ammonia in a very similar way to the reaction between (b) hydrogen chloride and ammonia.

REDOX REACTIONS

8.1 INTRODUCTION TO REDOX REACTIONS

- **Redox reactions** involve *red*uction and *ox*idation:
  - **Oxidation** is the loss of electrons
  - **Reduction** is the gain of electrons
  - A simple mnemonic is **OIL RIG** (oxidation is loss, reduction is gain).
- The **oxidizing agent (oxidant)** causes oxidation to occur: in the process, the oxidizing agent is itself reduced (and so gains electrons).
- The **reducing agent (reductant)** causes reduction to occur: in the process, the reducing agent is itself oxidized (and so loses electrons).

8.2 OXIDATION NUMBER

- The **oxidation number** of an element in a species is the number of electrons that need to be added to the element to make a neutral atom. There is no agreed symbol, so we will use Ox(E) to show the oxidation number of element E.
  - The terms 'oxidation number' and '**oxidation state**' are frequently used interchangeably.
  - The sum of the oxidation numbers must equal the charge on the species (zero in the case of a neutral compound).
  - For a simple ion, the oxidation number equals the charge on the ion.
  - We assign any shared electron pair completely to the more electronegative atom.
  - The oxidation numbers of two common elements are usually
    - oxygen Ox(O) = −2
    - hydrogen Ox(H) = +1
      - Taking the Next Step 8.1 explains some exceptions to this simple rule of thumb.
- During redox reactions, **oxidation numbers must change**.

Unusual oxidation numbers for oxygen and hydrogen

While oxygen, when combined in a compound, almost always has oxidation number -2, there are a few important exceptions. In the molecule **hydrogen peroxide** H_2O_2, the two oxygen atoms are bonded together and hence the electrons in that bond are shared equally. As each of the two hydrogens has $Ox(H) = +1$, the two oxygens must add up to -2 to make the total zero for this neutral molecule; hence each oxygen has $Ox(O) = -1$. (The same applies to compounds such as sodium peroxide Na_2O_2.) **Potassium super-oxide** KO_2 on the other hand has $Ox(K) = +1$ and, unusually, $Ox(O) = -1/2$.

In the compound **oxygen difluoride** OF_2, oxygen combines with the only element with a higher electronegativity and so has a *positive* oxidation number of $+2$ with $Ox(F) = -1$, as always for fluorine when it is in a compound.

Hydrogen forms compounds called **hydrides** with alkali metals such as sodium (sodium hydride is NaH). In hydrides, hydrogen is the *more* electronegative atom and hence has an oxidation number $Ox(H) = -1$.

8.3 SIMPLE HALF-EQUATIONS

- It is often very useful to break down the equation for a redox reaction into its constituent reduction and oxidation **half-equations**, each of which focuses on what is happening to just *one* element.
  - All half-equations must balance, both for atoms *and* for charge.
- **Common reducing agents** include the metals magnesium, aluminium, and zinc: in each case, the metal is oxidized to a positive ion (the number of electrons involved will be the same as the charge on the ion formed):
  - $Mg \rightarrow Mg^{2+} + 2e^-$
  - $Al \rightarrow Al^{3+} + 3e^-$
  - $Zn \rightarrow Zn^{2+} + 2e^-$
- **Common oxidizing agents** include oxygen and the halogens chlorine and bromine. In each case the non-metal is reduced to a negative ion:
  - $O_2 + 4e^- \rightarrow 2O^{2-}$
  - $Cl_2 + 2e^- \rightarrow 2Cl^-$
  - $Br_2 + 2e^- \rightarrow 2Br^-$
- To **combine two half-equations** to make a balanced chemical equation, we must make sure that **the number of electrons is balanced**.
  - When oxygen reacts with magnesium, we multiply the magnesium half-equation by 2 as it involves only 2 electrons while the oxygen half-equation involves 4 electrons:
    - $2Mg + O_2 \rightarrow 2MgO$ (we can consider MgO as containing one Mg^{2+} ion and one O^{2-} ion).

- When chlorine reacts with aluminium, we multiply the aluminium half-equation by 2 as it involves 3 electrons while multiplying the chlorine half-equation by 3 as it involves 2 electrons:
  - $2Al + 3Cl_2 \rightarrow 2AlCl_3$ (we can consider $AlCl_3$ as containing one Al^{3+} ion and three Cl^- ions)
  - The electrons always cancel when the full equation is formed.
- The reaction of **acids with metals is a *redox* reaction**, not an acid–base reaction.
  - When magnesium reacts with dilute acids, hydrogen gas is evolved and a solution of a magnesium salt is formed.
  - The half-equation describing the formation of hydrogen gas from an acid is
    - $2H^+ + 2e^- \rightarrow H_2$
  - Combining the half-equations $Mg \rightarrow Mg^{2+} + 2e^-$ and $2H^+ + 2e^- \rightarrow H_2$ gives the balanced equation (because the number of electrons involved is 2 in both cases):
    - $Mg + 2H^+ \rightarrow Mg^{2+} + H_2$

8.4 SOME HARDER HALF-EQUATIONS

- More complicated half-equations often involve **acid as a source of hydrogen ions**.
  - Two exceptionally important oxidizing agents in aqueous solution are **acidified manganate(VII) ion** and **acidified dichromate(VI) ion**:
    - $MnO_4^- + 8H^+ + 5e^- \rightarrow Mn^{2+} + 4H_2O$ Ox(Mn) changes from +7 to +2
    - $Cr_2O_7^{2-} + 14H^+ + 6e^- \rightarrow 2Cr^{3+} + 7H_2O$ Ox(Cr) changes from +6 to +3

    To see how to work out the manganate(VII) ion half-equation, note that Ox(Mn) falls from +7 to +2, so 5 electrons must have been added. The 4 oxygens in the manganate(VII) ion end up in 4 water molecules, so 8 hydrogen ions must be needed. As a final check, make sure that the charges balance (a total of 2+ on each side, in this case).
- As for the simpler examples above, to combine two half-equations to make a balanced chemical equation, we must make sure that the number of electrons is balanced. Manganate(VII) ions oxidize iron(II) ions to iron(III) ions. The half-equation for the oxidation of iron(II) ion to iron(III) ion is

$$Fe^{2+} \rightarrow Fe^{3+} + e^-$$

As the half-equation above involves only one electron, we need to multiply it by 5 to make the electrons balance with the manganate(VII) ion half-equation, giving

$$MnO_4^- + 8H^+ + 5Fe^{2+} \rightarrow 5Fe^{3+} + Mn^{2+} + 4H_2O$$

As a final check, make sure that the charges balance (a total of 17+ on each side).

8.5 ELECTROCHEMICAL CELLS

Zinc

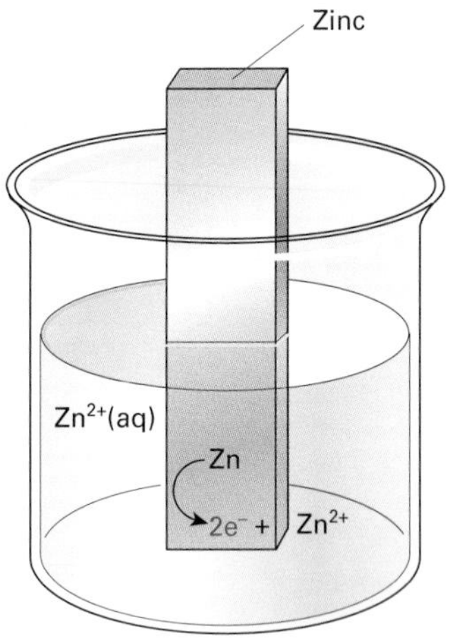

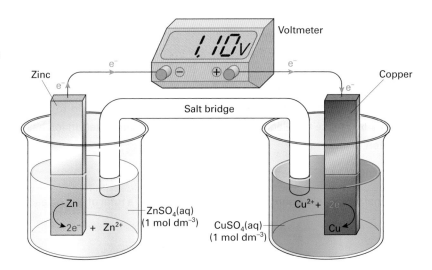

FIGURE 8.1 A zinc half-cell has zinc metal dipping into a solution containing zinc ions.

Occasionally, precipitation reactions might occur: for example, potassium chloride cannot be used if a silver half-cell is present.

- Copper metal does *not* react with dilute hydrochloric acid to form hydrogen gas (copper lies below hydrogen in the reactivity series); copper metal *does*, however, react with nitrate ion (from concentrated nitric acid for example) to form nitrogen dioxide.
  - To be able to make **predictions** about such behaviour, we must make measurements.
  - Measurements on electrochemical cells enable us to construct a list of standard electrode potentials (Table 8.1).
- We can make an **electrochemical cell** by connecting two **half-cells** together: the most common is a metal/metal-ion half-cell.
- A **metal/metal-ion half-cell** consists of a strip of metal dipping into a solution of one of the metal's ions. One example is a zinc half-cell, as shown in Figure 8.1.
  - If the concentration of the solution is **1 mol dm^{-3}** it is a **standard** zinc half-cell.
- We can connect two metal/metal-ion half-cells together using a **salt bridge** (typically containing potassium chloride or potassium nitrate) between the two solutions with a **digital voltmeter** between the clips attached to the metals, as shown in Figure 8.2.

  > A digital voltmeter has a very high internal resistance and so there is a negligible drop in potential across its terminals.

- A **gas half-cell** consists of a gas passing over a solid catalytic surface into a solution of its ions. Figure 8.3 involves the most important example: the standard hydrogen half-cell or **standard hydrogen electrode** (**SHE**), where
  - hydrogen gas (at 1 bar)
  - passes over a platinum electrode
  - into a solution of 1 mol dm^{-3} strong acid (i.e. at pH 0.0)
  - at 298 K.

FIGURE 8.2 A standard copper half-cell is connected to a standard zinc half-cell. The salt bridge connects the solutions together. The voltmeter displays the resulting standard cell potential.

- Taking the Next Step 8.2 discusses one type of half-cell which is less common pre-university.

The redox half-cell

Another type of half-cell (sometimes called a **redox half-cell**) can be made from a metal that has two common oxidation states in aqueous solution, such as iron(II) and iron(III) ions. There needs to be a conductor, usually platinum, present to conduct the current. The representation (Section 8.7) of this half-cell is

$$Pt(s) | Fe^{2+}(aq), Fe^{3+}(aq)$$

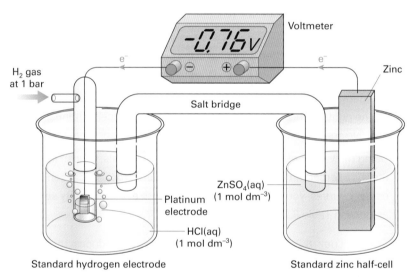

FIGURE 8.3 We can measure a standard electrode potential by connecting a standard half-cell (zinc, in this case) to a standard hydrogen electrode (*on the left*). The voltmeter displays the standard electrode potential.

8.6 STANDARD ELECTRODE POTENTIALS

- The **standard electrode potential** $E^{\ominus}$ for a given half-cell is equal to the **e.m.f.** (electromotive force) shown on the digital voltmeter when the standard half-cell is on the right and a SHE is on the left, as shown in Figure 8.3. Table 8.1 lists some representative values: they are all written as reductions.

- The **standard cell potential** (or **standard cell e.m.f.**) $E^{\ominus}$ **(cell)** is given by the equation:

  – $E^{\ominus}$ **(cell)** = $E^{\ominus}$ **(right-hand half-cell)** – $E^{\ominus}$ **(left-hand half-cell)**

    ○ For the reaction between zinc and copper(II) ions shown in Figure 8.2:
    $$E^{\ominus}(cell) = E^{\ominus}(Cu^{2+}/Cu) - E^{\ominus}(Zn^{2+}/Zn) = (+0.34\ V) - (-0.76\ V)$$
    $$= +1.10\ V$$

TABLE 8.1 Standard electrode potentials.

Oxidized species	⇌	Reduced species	$E^{\ominus}$/V
$K^+(aq) + e^-$	⇌	K(s)	−2.92
$Ca^{2+}(aq) + 2e^-$	⇌	Ca(s)	−2.87
$Mg^{2+}(aq) + 2e^-$	⇌	Mg(s)	−2.37
$Al^{3+}(aq) + 3e^-$	⇌	Al(s)	−1.66
$Zn^{2+}(aq) + 2e^-$	⇌	Zn(s)	−0.76
$Fe^{2+}(aq) + 2e^-$	⇌	Fe(s)	−0.44
$2H^+(aq) + 2e^-$	⇌	H_2(g)	0
$Cu^{2+}(aq) + 2e^-$	⇌	Cu(s)	+0.34
$Ag^+(aq) + e^-$	⇌	Ag(s)	+0.80

Although it is possible to predict whether a redox reaction is spontaneous, it is not possible to predict the *rate* at which the reaction occurs. A spontaneous reaction may be too slow to be of use commercially.

- If $E^{\ominus}$ **(cell) is positive**, the reaction taking place is **spontaneous** (Section 9.1).

  ○ Taking the Next Step 8.3 explains the connection between the standard cell potential $E^{\ominus}$ and the standard Gibbs energy change $\Delta G^{\ominus}$.

- It is good practice to place the half-cell with the more positive $E^{\ominus}$ on the right, as this will guarantee that the standard cell potential is positive and therefore that the cell reaction will be spontaneous.

- Using the values of their standard electrode potentials, we can list metals in order of their reactivity to form the **electrochemical series** (in earlier courses, you will probably have called this the **reactivity series**), as done in Table 8.1.

A Deeper Look 8.2 (at the end of this chapter) introduces the Nernst equation, which enables calculations when concentrations differ from 1 mol dm^{-3}.

TAKING THE NEXT STEP 8.3

What is the link between $E^{\ominus}$ and $\Delta G^{\ominus}$?

There is a direct proportionality between the standard cell potential $E^{\ominus}$ and the standard Gibbs energy change $\Delta G^{\ominus}$ (Section 9.3) for the reaction taking place in the cell, given by the equation:

$$\Delta G^{\ominus} = -zFE^{\ominus} \tag{1}$$

where z is the number of electrons involved in the reaction and F is the **Faraday constant**, which is the charge per mole of electrons (hence $F = N_A e$, which gives the value $F = 9.65 \times 10^4$ C mol^{-1}). Reactions are spontaneous if the standard Gibbs energy change is *negative* (Section 9.3) and therefore if the standard cell potential is *positive*. A Deeper Look 8.1 shows how we can use equation (1) above to find an unknown $E^{\ominus}$ value.

A DEEPER LOOK 8.1

How can we find an unknown $E^{\ominus}$ value?

Given the following $E^{\ominus}$ values,

(a) $Fe^{3+} + 3e^- \rightarrow Fe \qquad E^{\ominus} = -0.04$ V

(b) $Fe^{2+} + 2e^- \rightarrow Fe \qquad E^{\ominus} = -0.44$ V

we can find the $E^{\ominus}$ value for

(c) $Fe^{3+} + e^- \rightarrow Fe^{2+}$

We first need to convert the $E^{\ominus}$ values into $\Delta G^{\ominus}$ values using equation (1) above. We can then combine the standard Gibbs energy changes together (Section 9.3) in a similar way to Hess's law calculations. After combining the standard Gibbs energy changes, we can then use equation (1) again to reconvert to an $E^{\ominus}$ value.

$$\Delta G^{\ominus}(a) = -3FE^{\ominus} = -3F(-0.04 \text{ V}) = (+0.12 \text{ V})F$$
$$\Delta G^{\ominus}(b) = -2FE^{\ominus} = -2F(-0.44 \text{ V}) = (+0.88 \text{ V})F$$

Equation (c) ($Fe^{3+} + e^- \rightarrow Fe^{2+}$) is equal to (a) − (b), so we can find its standard Gibbs energy change by subtracting $\Delta G^{\ominus}(b)$ from $\Delta G^{\ominus}(a)$:

$$\Delta G^{\ominus}(c) = \Delta G^{\ominus}(a) - \Delta G^{\ominus}(b) = (+0.12 \text{ V})F - (+0.88 \text{ V})F = (-0.76 \text{ V})F$$

Finally, using equation (1) again, with $z = 1$,

$$\Delta G^{\ominus}(c) = -FE^{\ominus}(c) = (-0.76 \text{ V})F \text{ so } E^{\ominus}(c) = +0.76 \text{ V}$$

This roundabout method is needed as the number of electrons in reactions (a) and (b) is different. Notice that it is convenient to carry the Faraday constant F through the calculation, as it will cancel in the end.

8.7 CELL REPRESENTATIONS

There is a convention for representing the cell formed by the combination of two half-cells. We represent the cell shown in Figure 8.2 as follows:

$$Zn(s)|Zn^{2+}(aq)||Cu^{2+}(aq)|Cu(s)$$

This convention means that

- we place the metal electrodes through which the electrons flow at opposite ends,
- each vertical line separates different phases, for example the solid from the aqueous solution,
- the double vertical line represents the salt bridge,
- the standard cell potential is
$$-E^{\ominus}(\text{cell}) = E^{\ominus}(Cu^{2+}/Cu) - E^{\ominus}(Zn^{2+}/Zn) = (+0.34 \text{ V}) - (-0.76 \text{ V})$$
$$= +1.10 \text{ V}$$

- and hence the forward reaction shown below is spontaneous

  $\circ$ $(Cu^{2+} + 2e^- \rightleftharpoons Cu) - (Zn^{2+} + 2e^- \rightleftharpoons Zn)$ or $Zn + Cu^{2+} \rightarrow Zn^{2+} + Cu$

This reaction is the basis of the Daniell cell; see Taking the Next Step 8.4. We represent the cell shown in Figure 8.3 as follows:

$$Pt(s)|H_2(g)|H^+(aq) \| Zn^{2+}(aq)|Zn(s)$$

- The standard cell potential is

  - $E^{\ominus}(\text{cell}) = E^{\ominus}(Zn^{2+}/Zn) - E^{\ominus}(H^+/H_2) = (-0.76 \text{ V}) - (0 \text{ V}) = -0.76 \text{ V}$

  - and hence the forward reaction shown below is **not** spontaneous

    $\circ$ $(Zn^{2+} + 2e^- \rightleftharpoons Zn) - (2H^+ + 2e^- \rightleftharpoons H_2)$ or $Zn^{2+} + H_2 \rightarrow Zn + 2H^+$

  - The spontaneous reaction will therefore be the *backward* reaction, namely

    $\circ$ $Zn + 2H^+ \rightarrow Zn^{2+} + H_2$

TAKING THE NEXT STEP 8.4

Galvanic cells and electrolytic cells; cathodes and anodes

Galvanic cells (or **voltaic cells**) use chemical reactions to generate electricity. The first reliable example was the Daniell cell, based on the spontaneous reaction between zinc and copper(II) ions shown in Figure 8.2. We call the Daniell cell and the common alkaline dry cell **non-rechargeable cells** (or primary cells). **Rechargeable cells** (or secondary cells) include nickel-cadmium, lead-acid, and lithium-ion cells. **Fuel cells** can generate electricity as long as we provide a supply of fuel, such as hydrogen.

Electrolytic cells on the other hand use electricity to cause chemical change. Using the standard electrode potentials for the two half-cells, we can work out the standard cell potential. If a voltage *larger* than this is applied in the opposite direction, it is possible to drive a reaction in its non-spontaneous direction. So we can electrolyse molten sodium chloride to produce sodium metal and chlorine gas. Figure 8.4 shows the Downs

FIGURE 8.4 The Downs cell manufactures sodium metal by electrolysis.

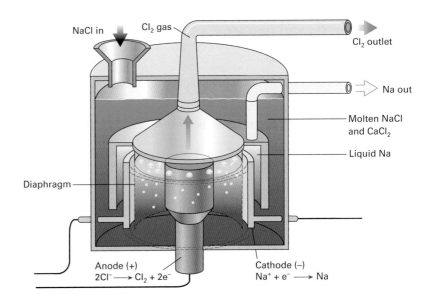

cell, which extracts sodium metal from molten sodium chloride industrially. (Electrolysing *aqueous* sodium chloride introduces the complication that water can be decomposed instead of chloride ions being discharged.)

When you first met the terms 'anode' and 'cathode' you were probably studying electrolysis, and the charges on the electrodes will have been stated. However, there is a superior way of explaining these terms, which applies to **both** galvanic cells and electrolytic cells:

Anode: the site of oxidation

Cathode: the site of reduction

An easy mnemonic is **A**node/oxid**A**tion and **C**athode/redu**C**tion.

A simplified version of the oxidation reaction occurring at the **anode** in a **lithium-ion cell** is

$$Li \rightarrow Li^+ + e^-$$

in which lithium is oxidized from oxidation number 0 to +1.

A more realistic version recognizes the role played by the graphite electrode:

$$LiC_6 \rightarrow Li^+ + e^- + 6C$$

The structure of LiC_6 involves lithium inserted (**intercalated**) between the layers of graphite.

The reduction reaction occurring at the **cathode** in a typical lithium-ion cell is

$$CoO_2 + Li^+ + e^- \rightarrow LiCoO_2$$

in which cobalt is reduced from oxidation number +4 to +3.

Returning to electrolysis, the anodic reaction in the Downs cell, shown in Figure 8.4, is

$$2Cl^- \rightarrow Cl_2 + 2e^- \qquad \text{Ox(Cl) changes from } -1 \text{ to } 0$$

The cathodic reaction is

$$Na^+ + e^- \rightarrow Na \qquad \text{Ox(Na) changes from } +1 \text{ to } 0$$

The **anion** Cl^- is attracted to the anode, which is *positive in an electrolytic cell*. The **cation** Na^+ is attracted to the cathode, which is *negative in an electrolytic cell*.

Because electrons flow towards the cathode in a galvanic cell where they are used to cause reduction, the *cathode is positive in a galvanic cell*: **note the opposite polarity**.

A DEEPER LOOK 8.2

What is the Nernst equation?

The Nernst equation allows us to make quantitative predictions about what happens when concentrations differ from the standard value of 1 mol dm^{-3}.

Pre-university, you are likely to have ventured very cautiously into the situation where concentrations differ from 1 mol dm^{-3}, and may have had rather vague warnings of the effect on predictions.

The equation Walther Nernst introduced allows for 'fine-tuning' the actual cell potential E around the value of the standard cell potential $E^\ominus$, as a result of the fact that the concentrations of the solutions may not be 1 mol dm^{-3}. A simplified version of the Nernst equation is

$$E = E^\ominus + \frac{0.059\text{ V}}{z}\log_{10}\left(\frac{[\text{oxidized species}]}{[\text{reduced species}]}\right)$$

where z is the number of electrons involved in the redox equation.

An important application is to nerve cells, where a potential difference can build up because of different concentrations of potassium ions inside and outside the cell membrane (the potassium ion concentration is about 25 times larger inside); $z = 1$, as a potassium ion needs only one electron to become the metal:

$$E = (0.059\text{ V})\log_{10}\left(\frac{[\text{K}^+, \text{outside}]}{[\text{K}^+, \text{inside}]}\right) = -(0.059\text{ V})\log_{10} 25 = -82\text{ mV}$$

which is reasonably close to the measured value of -70 mV (the discrepancy is greatly reduced if we also include the effect of the sodium ion imbalance).

Another important application is to reactions that depend significantly on the hydrogen ion concentration, as this can easily be changed by 14 orders of magnitude by going from 1 mol dm^{-3} strong acid to 1 mol dm^{-3} strong base. $E^\ominus$ *values are measured at pH 0*, because the standard hydrogen electrode has $[\text{H}^+] = 1$ mol dm^{-3}.

An important application is to the reduction of manganate(VII) ion:

$$\text{MnO}_4^- + 8\text{H}^+ + 5e^- \rightarrow \text{Mn}^{2+} + 4\text{H}_2\text{O} \qquad E^\ominus = +1.51\text{V}$$

At a pH other than zero, the following equation applies:

$$E = E^\ominus + \frac{0.059\text{ V}}{5}\log_{10} [\text{H}^+]^8 = E^\ominus + \frac{8 \times 0.059\text{ V}}{5}\log_{10} [\text{H}^+]$$

$$= E^\ominus - \frac{8 \times 0.059\text{ V}}{5}\text{pH}$$

At pH 14, $E = +1.51\text{V} - \dfrac{8 \times 0.059\text{ V}}{5} \times 14 = +0.19$ V. The reduction of manganate(VII) ion to manganese(II) ion is strongly favoured at pH 0 but *not* at pH 14. An alternative reduction product is manganate(VI) ion, which is coloured green in aqueous solution:

$$\text{MnO}_4^- + e^- \rightarrow \text{MnO}_4^{2-} \qquad E^\ominus = +0.56\text{ V}$$

Because no hydrogen ions are involved in the reaction above, the electrode potential is the *same* at both pH 0 and pH 14. In acidic solution, manganate(VI) ion is not formed, but it *is* formed in alkaline solution. A simple colour test for alkenes involves adding the suspected alkene to *alkaline* potassium manganate(VII) and observing the purple solution change to green, as illustrated in Figure 8.5.

FIGURE 8.5 In alkaline solution, the alkene cyclohexene can reduce purple manganate(VII) ions to green manganate(VI) ions.

Potassium manganate(VII) is often called potassium permanganate at university (Section 14.2).

SPONTANEOUS CHANGE, ENTROPY, AND GIBBS ENERGY

9.1 SPONTANEOUS CHANGE

- **Spontaneous changes have a natural tendency to occur**; you may have come across the word **'feasible'** used instead of 'spontaneous'.
  - 'Spontaneous' when used scientifically does not necessarily mean fast. Gibbs energy changes (Section 9.3) cannot be used to predict the *rate* of the reaction.
- By the end of the nineteenth century, scientists noticed that a very large majority of the reactions that happened were exothermic (Section 5.1). The challenge was then to explain why spontaneous *endothermic* reactions happen.
- The missing concept needed to explain spontaneous endothermic reactions was first introduced into science by Rudolf Clausius in 1865: **entropy**.

Entropy

- **Entropy is a measure of the dispersal of energy**.
  - The more the energy in a system is spread out, the higher its entropy is.
- We define the **entropy change** in a system as
  - $\Delta S = q/T$
    - where q is the heat added to the system, and T is the thermodynamic temperature (in kelvin) of the system.
- The entropy of a substance increases as the temperature increases: the largest change occurs when a liquid changes to a gas as the particles and the energy they carry become much more spread out.
- Another common description of entropy is that it is a **measure of disorder**: increased disorder causes increased entropy. So gases generally have much higher entropy than liquids, as shown in Figure 9.1.

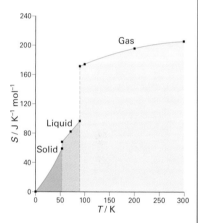

FIGURE 9.1 The entropy of oxygen as a function of temperature. Note the large increase from liquid to gas.

This equation only strictly applies to **'reversible'** heat transfers. A reversible transfer ensures that no hot spots form in the system that later disperse spontaneously, hence further increasing the entropy.

What is the criterion for spontaneous change?

- The **Second Law of thermodynamics** states that the total entropy of the universe always tends to increase.
  - The word 'tends' means that we know the *direction* of change but not how fast it will occur.
  - The total entropy *can* remain the same.
- The **total entropy of the universe** is the sum of the entropy change in the **system** (which we are studying) and the entropy change in the **surroundings** (which is everything else).
  - We can work out the entropy change in the surroundings because we know that the heat added to the surroundings must be equal in magnitude but opposite in sign to the heat added to the system (by the First Law of thermodynamics, Section 5.2).
    - The heat added to the system (at constant pressure) is the enthalpy change ΔH (Section 5.1), so the heat added to the surroundings is $-\Delta H$.
- Hence the **entropy change in the surroundings** ΔS(surr.) is equal to $-\Delta H/T$, where T is the temperature of the surroundings in this case.
  - So the criterion for spontaneous change is that
    - ΔS(system) $+ \Delta S$(surr.) > 0
    - **Spontaneous change can only happen when $\Delta S - \Delta H/T > 0$**
- Note that the values of both ΔS and ΔH are those for the **system**: we now only need to concentrate on the system.

Simple examples of the general theory

- If the **entropy change in the system is negligible**, the only reactions that can occur are those that are exothermic: ΔH *must* be negative so that $-\Delta H/T$ is positive.
  - This is the reason why so many spontaneous reactions are exothermic (as illustrated in Figure 9.2): the energy spread out during the reaction causes a positive entropy change in the surroundings.
- If the **enthalpy change in the system is negligible**, the only processes that can occur are those in which the entropy of the system increases.
  - The simplest example is the spreading out of particles during **diffusion**.
    - There is no enthalpy change, but the entropy of the system increases as the particles spread out.
  - Another simple example is **mixing**: the mixed state is more disordered and so has higher entropy.
- Now we can answer the question about **spontaneous endothermic reactions**.
  - First, they can only occur if *the entropy change in the system is positive*, typically due to the release of a gas.

FIGURE 9.2 The Space Shuttle.

- Second, they can only occur *above a certain temperature*: T must be large enough that ΔS has a larger value than $\Delta H/T$.
  - ○ Calcium carbonate is stable to decomposition at room temperature (the White Cliffs of Dover have not disappeared in the last 50,000 years), despite the positive entropy change in the system caused by the release of carbon dioxide gas. However, in a blast furnace the temperature T *is* high enough that ΔS is larger than $\Delta H/T$.
  - ○ We can find the temperature at which the reaction first becomes spontaneous using $T = \Delta H/\Delta S$
- We can turn these simple qualitative arguments into exact calculations.

9.2 STANDARD ENTROPY CHANGE

- We define standard entropy changes in *exactly the same way* as standard enthalpy changes (Section 5.1).
- The **standard entropy change $\Delta S^{\ominus}$** (sometimes written $\Delta_r S^{\ominus}$) is the
  - entropy change per mole of the reaction as written
  - when reactants in their standard states form products in their standard states
  - at 1 bar

  You may have been told that the standard entropy change is measured at 298 K. In fact, 298 K is simply the *most common* temperature at which the values are measured, when the symbol should be written as $\Delta S^{\ominus}(298\ \text{K})$.

- We can find standard entropy changes in *exactly the same way* that we can find standard enthalpy changes using Hess's law (Section 5.2).

You may have come across the expression 'entropy change when one mole . . .'. This gets the units wrong, as it would give a value in J K^{-1} rather than J K^{-1} mol^{-1}.

There is a similar equation to find the standard entropy change for a reaction:

- $\Delta S^{\ominus} = \Sigma S^{\ominus}(\text{products}) - \Sigma S^{\ominus}(\text{reactants})$.
  - Remember that the summation sign (Σ) indicates that we need to multiply by the mole ratios for each substance (Section 3.3).

Example

Let us consider the ammonia synthesis:

$$N_2(g) + 3H_2(g) \rightarrow 2NH_3(g),$$

$$\Delta S^{\ominus} = 2S^{\ominus}(NH_3, g) - S^{\ominus}(N_2, g) - 3S^{\ominus}(H_2, g)$$

$$= 2(192\,J\,K^{-1}\,mol^{-1}) - (192\,J\,K^{-1}\,mol^{-1}) - 3(131\,J\,K^{-1}\,mol^{-1})$$

$$= -201\,J\,K^{-1}\,mol^{-1}$$

The value is negative, as there are fewer moles of gas on the right-hand side of the equation.

- One technical point is that standard entropies are *absolute* values; see A Deeper Look 9.1.

A DEEPER LOOK 9.1

The Third Law of thermodynamics

We can trace the reason why standard entropies are absolute values back to the **Third Law of thermodynamics**, which states that

The entropy of a perfectly crystalline substance at absolute zero is zero.

A perfect crystal has all the constituent particles in their correct places, so there is no disorder at all and thus its entropy is zero. The fact that measurements start from zero means that there is no delta in the symbol for the standard entropy ($S^{\ominus}$), unlike the case for the standard enthalpy change of formation ($\Delta_f H^{\ominus}$).

- It is important to stress that it is *no more difficult* to find the standard entropy change for a reaction than it is to find the standard enthalpy change for the same reaction.

Example

For calcium carbonate decomposition: $CaCO_3(s) \rightarrow CaO(s) + CO_2(g)$

- $\Delta H^{\ominus} = \Delta_f H^{\ominus}(CaO, s) + \Delta_f H^{\ominus}(CO_2, g) - \Delta_f H^{\ominus}(CaCO_3, s)$

$$= (-635\,kJ\,mol^{-1}) + (-394\,kJ\,mol^{-1}) - (-1207\,kJ\,mol^{-1}) = +178\,kJ\,mol^{-1}$$

- $\Delta S^{\ominus} = S^{\ominus}(CaO,s) + S^{\ominus}(CO_2,g) - S^{\ominus}(CaCO_3,s)$

  $= (40\,JK^{-1}\,mol^{-1}) + (214\,JK^{-1}\,mol^{-1}) - (93\,JK^{-1}\,mol^{-1})$

  $= +161\,JK^{-1}\,mol^{-1}$

The relatively high value for CO_2 gas makes the standard entropy change positive.

At a temperature of 298 K, $\Delta S^{\ominus} - \Delta H^{\ominus}/T = (+161\,JK^{-1}\,mol^{-1}) - (+178\,000\,J\,mol^{-1})/(298\,K) = -436\,JK^{-1}\,mol^{-1}$. The negative sign shows that the decomposition is **not** spontaneous at 298 K.

- We can calculate the temperature at which decomposition starts as follows:

  $T = \Delta H^{\ominus}/\Delta S^{\ominus}$

  $- T = \dfrac{+178\,kJ\,mol^{-1}}{+161\,JK^{-1}\,mol^{-1}} = \dfrac{+178\,000\,J\,mol^{-1}}{+161\,JK^{-1}\,mol^{-1}}$

  ○ Because the top line involves kJ in the units and the bottom line only J, we need to use 1 kJ = 1000 J: we can then cancel J top and bottom.

  − $T = 1110\,K$ (3 sf)

- We can apply these ideas equally well to *physical* processes such as boiling water.

  − $\Delta_{vap}H^{\ominus}$ at 298 K is +44.0 kJ mol^{-1} (boiling water is an endothermic process).

  − $\Delta_{vap}S^{\ominus}$ at 298 K is +119 J K^{-1} mol^{-1} (the entropy increases greatly on going from a liquid to a gas).

- Hence we can calculate the transition temperature (water's boiling point) as follows:

  − $T = \Delta_{vap}H^{\ominus}/\Delta_{vap}S^{\ominus}$

  − $T = \dfrac{+44.0\,kJ\,mol^{-1}}{+119\,JK^{-1}\,mol^{-1}} = \dfrac{+44\,000\,J\,mol^{-1}}{+119\,JK^{-1}\,mol^{-1}}$

  ○ Because the top line involves kJ in the units and the bottom line only J, we need to use 1 kJ = 1000 J: we can then cancel J top and bottom.

  − $T = 370\,K$ (3 sf)

  ○ The calculated value is within 1% of the true value (373 K).

An even closer match would be obtained if we were to repeat the calculation using the values measured at 370 K.

9.3 GIBBS ENERGY AND STANDARD GIBBS ENERGY CHANGE

- The first person to explain thermodynamics in complete detail was the American Willard Gibbs in the 1870s. Gibbs combined enthalpy and

The formal definition of Gibbs energy is
G = H − TS
So $\Delta G = \Delta H - T\Delta S - S\Delta T$
which, at constant temperature, gives the equation alongside.

entropy together to form a single physical quantity we now call the **Gibbs energy** (or **Gibbs free energy**).

• At constant temperature, the **Gibbs energy change ΔG** for a reaction is equal to the enthalpy change ΔH minus the entropy change ΔS multiplied by the thermodynamic temperature T of the system: $\Delta\boldsymbol{G} = \Delta\boldsymbol{H} - \boldsymbol{T}\Delta\boldsymbol{S}$.

• We define standard Gibbs energy changes in *exactly the same way* as standard enthalpy changes (Section 5.1).

• The **standard Gibbs energy change $\Delta\boldsymbol{G}^{\ominus}$** (or sometimes $\Delta_r\boldsymbol{G}^{\ominus}$) is the

  – Gibbs energy change per mole of the reaction as written

  – when reactants in their standard states form products in their standard states

  – at 1 bar

    ◦ Remember that the **standard state** of a substance is its pure form at 1 bar (Section 5.1).

You may come across the phrase 'under standard conditions' being used. However, this phrase fails to highlight the importance of the concept of standard states. Taking the Next Step 9.1 explains why this is so important.

Figure 9.7 shows very clearly that standard Gibbs energy changes vary with temperature.

You may have been told that the standard Gibbs energy change is measured at 298 K. In fact 298 K is simply the *most common* temperature at which the values are measured, when the symbol should be written as $\Delta G^{\ominus}$(298 K). Taking the Next Step 9.1 explains why this point is particularly important by connecting the standard Gibbs energy change to the equilibrium constant.

• We can find standard Gibbs energy changes in *exactly the same way* that we can find standard enthalpy changes using Hess's law (Section 5.2). There is a similar equation to find the standard Gibbs energy change for a reaction:

  – $\Delta G^{\ominus} = \Sigma\, \Delta_f G^{\ominus}(\text{products}) - \Sigma\, \Delta_f G^{\ominus}(\text{reactants})$

    ◦ Remember that the summation sign (Σ) indicates that it is necessary to multiply by the mole ratios for each substance (Section 3.3).

Example

For the ammonia synthesis $N_2(g) + 3H_2(g) \rightarrow 2NH_3(g)$,

$$\Delta G^{\ominus} = 2\Delta_f G^{\ominus}(NH_3, g) - \Delta_f G^{\ominus}(N_2, g) - 3\Delta_f G^{\ominus}(H_2, g)$$
$$= 2(-16.5\,\text{kJ mol}^{-1}) - (0\,\text{kJ mol}^{-1}) - 3(0\,\text{kJ mol}^{-1})$$
$$= -33.0\,\text{kJ mol}^{-1}$$

• The value for $\Delta_f G^{\ominus}$ for an element is zero by definition.

• The standard Gibbs energy change is related to the standard enthalpy and standard entropy changes by the equation

$$\Delta\boldsymbol{G}^{\ominus} = \Delta\boldsymbol{H}^{\ominus} - \boldsymbol{T}\Delta\boldsymbol{S}^{\ominus}$$

• Spontaneous change can only happen when $\Delta S^{\ominus} - \Delta H^{\ominus}/T > 0$ (Section 9.1).

– Dividing the equation for the standard Gibbs energy change by $-T$ gives

$$-\Delta G^{\ominus}/T = \Delta S^{\ominus} - \Delta H^{\ominus}/T$$

> You may have come across the Gibbs energy change described as a *balance* between the enthalpy change in the system and the entropy change in the system. 'Balance' is a poor choice of word; the enthalpy change is directly related to the entropy change in the surroundings, so the Gibbs energy change is directly related to the total entropy change in the universe (albeit with the opposite sign). The *only* criterion for change is that the total entropy of the universe increases (Section 9.1), which corresponds to a negative standard Gibbs energy change.

• Because the thermodynamic temperature T cannot be negative, we can conclude that **spontaneous change can only happen when $\Delta G^{\ominus} < 0$**.

– Setting $T = \Delta H^{\ominus}/\Delta S^{\ominus}$ (Section 9.2) found the *lowest* temperature at which $\Delta G^{\ominus}$ becomes negative and hence an endothermic reaction becomes spontaneous. In fact, the discussion is a little more complex: Taking the Next Step 9.1 explains why.

Example

Using the following two combustion reactions, we can find $\Delta G^{\ominus}$ for the conversion of diamond to graphite:

$$C(s, diamond) + O_2(g) \rightarrow CO_2(g) \qquad \Delta_c G^{\ominus} = -395.4 \, kJ \, mol^{-1}$$
$$C(s, graphite) + O_2(g) \rightarrow CO_2(g) \qquad \Delta_c G^{\ominus} = -393.5 \, kJ \, mol^{-1}$$

We can use a Hess's law cycle (Section 5.2):

$$C(s, diamond) \rightarrow C(s, graphite)$$
$$\searrow \qquad \swarrow$$
$$CO_2(g)$$

to find that $\Delta G^{\ominus} = (-395.4 \, kJ \, mol^{-1}) - (-393.5 \, kJ \, mol^{-1}) = -1.9 \, kJ \, mol^{-1}$

The negative value for $\Delta G^{\ominus}$ tells us that the conversion of diamond to graphite is spontaneous.

• One final point needs emphasizing carefully: using standard Gibbs energy changes, **we can find no information about the rate of reaction**. A reaction with a negative standard Gibbs energy change *will* happen, but it may be incredibly slow. The classic example is that diamond is thermodynamically unstable relative to graphite (see example immediately above), but jewellers and spouses need not worry: diamond rings decompose *exceptionally* slowly.

TAKING THE NEXT STEP 9.1

Standard Gibbs energy change and equilibrium

Using the standard Gibbs energy change would seem to divide all reactions into two groups:

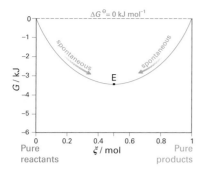

FIGURE 9.3 The graph of Gibbs energy G vs extent of reaction ξ (defined immediately below right) for a reaction $A + B \rightleftharpoons C + D$, where $\Delta G^{\ominus} = 0\,\text{kJ mol}^{-1}$. The Gibbs energy of the pure reactants is the same as that of the pure products. Reaction spontaneously proceeds from either direction towards equilibrium (E) because the Gibbs energy change of mixing of reactants and products is negative. The largest Gibbs energy change of mixing occurs when each substance is present in equal amount.

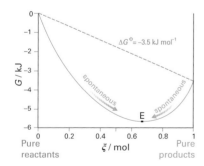

FIGURE 9.4 The graph of Gibbs energy G vs extent of reaction ξ for the formation of ethyl ethanoate. An extent of reaction of 0.67 mol corresponds to an equilibrium constant of $\dfrac{(0.67\,\text{mol}) \times (0.67\,\text{mol})}{(0.33\,\text{mol}) \times (0.33\,\text{mol})} = 4$

- Reactions that have a negative $\Delta G^{\ominus}$ value are spontaneous (Section 9.1).
- Reactions that have a positive $\Delta G^{\ominus}$ value are *not* spontaneous.

This all-or-nothing split appears not to link at all with the idea of chemical equilibrium (Section 6.1), where we know that reactions produce a mixture of reactants and products.

We stressed previously that it is very important that the **standard** Gibbs energy change concerns change from *pure* reactants to *pure* products: this is why the concept of the **standard state** (the pure substance at 1 bar) is so important. At no point until right now has the idea of *mixing* reactants and products been included. But why is this idea so important to us? **Mixing always generates a positive entropy change** (Section 9.1), so this *additional* entropy change needs to be considered because it will cause a **negative Gibbs energy change of mixing**.

Let us consider a hypothetical reaction with *zero* standard Gibbs energy change. Some reaction will happen despite the standard Gibbs energy change being zero, driven by the negative Gibbs energy change caused by mixing some products with the reactants. Reaction will occur until the Gibbs energy is a minimum. To simplify the explanation, **we will concentrate on a reaction with the equation $A + B \rightleftharpoons C + D$**. The Gibbs energy change of mixing is largest when each substance is present in equal amount, i.e. half-way to completion, as represented by point E on the graph in Figure 9.3.

To proceed further we need to introduce a quantity unfamiliar from pre-university courses: the extent of reaction. We define the **extent of reaction ξ** (the Greek letter xi) so that pure reactants correspond to a ξ value of 0 mol and pure products to a ξ value of 1 mol.

When $\Delta G^{\ominus} = 0\,\text{kJ mol}^{-1}$, equilibrium occurs when $\xi = 0.5\,\text{mol}$, as we saw in Figure 9.3. There is a 50:50 mixture of reactants and products at equilibrium. But what happens when equilibrium is reached with products and reactants present in different proportions—that is, when the equilibrium lies towards products or reactants, rather than being half-way between the two?

For a reaction that has a *negative* standard Gibbs energy change (which we call an **exergonic** reaction), the products are favoured: at the minimum, ξ is greater than 0.5 mol. The Gibbs energy change of mixing (which depends only on the amount of each substance and so has the same value at any particular ξ as in Figure 9.3) remains present and ensures that a *minimum* still occurs in the Gibbs energy graph. However, now the minimum lies closer to the products than to the reactants, as we see in Figure 9.4.

The example shown in Figure 9.4 applies to the formation of ethyl ethanoate, a common equilibrium reaction (Section 6.1).

If the standard Gibbs energy change is *negative and large*, the line joining reactants and products falls very steeply; see Figure 9.5. At first glance, it may seem that the graph has no minimum. However, this is not the case, because the Gibbs energy change of mixing remains present. The minimum (which, we must remember, represents the position of equilibrium) now lies very far over towards the products: at the minimum, ξ has a value very close to 1 mol. Effectively 'the reaction goes to completion', as we see in Figure 9.5.

For a reaction that has a ***positive*** standard Gibbs energy change (which we call an **endergonic** reaction), the reactants are favoured: at the minimum, ξ is smaller than 0.5 mol. The Gibbs energy change of mixing remains present and ensures that a *minimum* still occurs in the Gibbs energy graph. However, now the minimum lies closer to the reactants than to the products.

If the standard Gibbs energy change is *positive and large*, the line joining reactants and products rises very steeply. Again, at first glance, it may seem that the graph has no minimum. However, this is not the case, because the Gibbs energy change of mixing remains present. The minimum now lies very far over towards the reactants: at the minimum, ξ has a value very close to 0 mol. Effectively 'no reaction occurs', as we see in Figure 9.6.

The larger the value of the standard Gibbs energy change $\Delta G^{\ominus}$, the further away the equilibrium will lie from a 50:50 mixture. The thermodynamic equilibrium constant K for the reaction (Section 6.1) describes the ratio of the products to the reactants at equilibrium. Therefore, there is clearly a connection between the standard Gibbs energy change $\Delta G^{\ominus}$ (the gradient of the line joining pure reactants and pure products) for a reaction and the reaction's equilibrium constant K.

The following equation (sometimes called the **reaction isotherm**) connects the two concepts:

$$\Delta G^{\ominus} = -RT \ln K \tag{1}$$

where R is the gas constant (Section 3.5) and T is the thermodynamic temperature; the symbol ln denotes the natural logarithm.

Now we can find the equilibrium constant for *any* reaction directly from the ***standard* Gibbs energy change** for the reaction.

When $\Delta G^{\ominus} = 0 \, \text{kJ mol}^{-1}$, there is a 50:50 mixture of reactants and products at equilibrium and the equilibrium constant $K = 1$. This agrees with the reaction isotherm because $\ln 1 = 0$.

For $\Delta G^{\ominus} = -3.5 \, \text{kJ mol}^{-1}$, the reaction isotherm shows that at $T = 298 \, \text{K}$,

$$-3500 \, \text{J mol}^{-1} = -(8.314 \, \text{J K}^{-1} \text{mol}^{-1})(298 \, \text{K}) \ln K$$

Hence $\ln K = 1.4$ and $K = e^{1.4} = 4$ (1 sf)

as, for example, for the formation of ethyl ethanoate; see Figure 9.4.

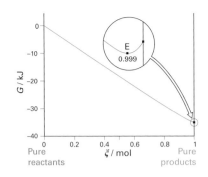

FIGURE 9.5 A reaction with a standard Gibbs energy change of −35 kJ mol^{-1} reaches equilibrium 99.9% of the way to completion.

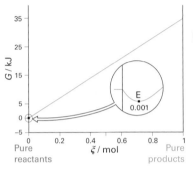

FIGURE 9.6 A reaction with a standard Gibbs energy change of +35 kJ mol^{-1} reaches equilibrium only 0.1% of the way to completion.

We can derive this equation by finding the minimum of the graph in the usual way: set the differential (in this case, $dG/d\xi$) equal to zero.

The reaction isotherm allows us to make some general predictions about the position of equilibrium for different values of $\Delta G^{\ominus}$. Table 9.1 summarizes these predictions:

TABLE 9.1 Standard Gibbs energy change connects to equilibrium position.

Standard Gibbs energy change, $\Delta G^{\ominus}$/kJ mol^{-1}	Equilibrium constant K	Equilibrium position
More negative than −35	Greater than 10^6	'Reaction effectively complete'
Between −35 and 0	Between 10^6 and 1	Products predominate
Between 0 and +35	Between 1 and 10^{-6}	Reactants predominate
More positive than +35	Smaller than 10^{-6}	'Effectively no reaction'

The reaction isotherm specifically involves the temperature on the right-hand side of the equation. Therefore, the value of the standard Gibbs energy change must be measured *at that particular temperature*. Sometimes this fact is emphasized at university by showing an explicit dependence of the standard Gibbs energy change on temperature:

$$\Delta G^{\ominus}(T) = -RT \ln K$$

Hence the importance stressed in Sections 5.1 and 9.2 that thermodynamic quantities should refer to a particular **stated** temperature.

A Deeper Look 9.2 shows how we can work out the effect of temperature on the equilibrium constant using the reaction isotherm.

A DEEPER LOOK 9.2

What is the effect of temperature on the equilibrium constant?

Armed with the reaction isotherm (equation (1) above), we can calculate how the equilibrium constant changes with temperature.

We can rearrange $\Delta G^{\ominus} = -RT \ln K$ to give the form

$$\ln K = -\Delta G^{\ominus}/RT$$

Then if we substitute $\Delta G^{\ominus} = \Delta H^{\ominus} - T\Delta S^{\ominus}$ (Section 9.3) into the above equation we find

$$\ln K = -\frac{\Delta H^{\ominus}}{RT} + \frac{\Delta S^{\ominus}}{R}$$

If we make the simplifying assumption that the standard enthalpy and standard entropy changes are constant over the temperature range chosen, then when we differentiate the equation above with respect to temperature, the only term that contributes is the one involving $\frac{1}{T}$: $\frac{1}{T}$ differentiates to $-\frac{1}{T^2}$

So $\dfrac{d \ln K}{dT} = \dfrac{\Delta H^{\ominus}}{RT^2}$

We call this the **van 't Hoff equation**. As the denominator on the right-hand side cannot be negative, the sign of the change in the natural logarithm of the equilibrium constant depends only on the sign of the standard enthalpy change. An endothermic reaction (for which $\Delta H^\ominus$ is positive) has an equilibrium constant that increases with temperature, just as Le Chatelier's principle states (Section 6.3). However, the van 't Hoff equation allows *quantitative* calculations to be done.

Jacobus van 't Hoff was the first recipient of the Nobel Prize in Chemistry (for his work on 'the laws of chemical dynamics and osmotic pressure in solutions').

A Deeper Look 9.3 introduces Ellingham diagrams.

A DEEPER LOOK 9.3

Ellingham diagrams

An **Ellingham diagram**, as shown in Figure 9.7, consists of graphs of the Gibbs energy change of formation $\Delta_f G^\ominus$ for various metal oxides as a function of temperature. (In the diagram, the $\Delta_f G^\ominus$ values are all given per mole

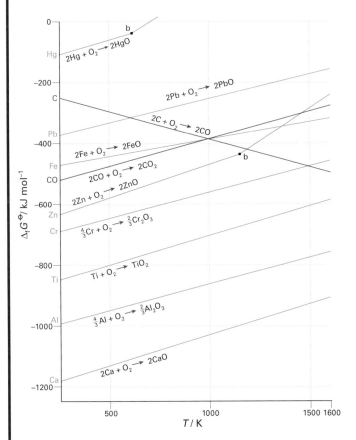

FIGURE 9.7 An Ellingham diagram showing the relationship between standard Gibbs energy change and temperature for the oxides of carbon and various metal oxides. For zinc, the slope becomes significantly greater at the boiling point (b).

Ellingham diagrams show very clearly that standard Gibbs energy changes depend on temperature (see Section 9.3).

of oxygen gas.) As $\Delta_f G^\ominus = \Delta_f H^\ominus - T\Delta_f S^\ominus$, the gradient is $-\Delta_f S^\ominus$ (assuming that $\Delta_f H^\ominus$ and $\Delta_f S^\ominus$ are constant).

The gradients of the metal/metal oxide plots are positive, as one mole of oxygen *gas* is destroyed in making the oxide, making $\Delta_f S^\ominus$ negative. Once the $\Delta_f G^\ominus$ value becomes positive, the oxide is no longer stable: decomposition happens for mercury(II) oxide, HgO, at about 700 K. Historically, heating mercury(II) oxide was the method Joseph Priestley used to discover oxygen gas (he heated the oxide using sunlight focused with a magnifying glass!).

Ellingham diagrams also contain plots for the formation of the oxides of carbon. The line for the conversion of carbon to carbon monoxide slopes *downwards*, because two moles of CO gas are produced for each mole of oxygen gas. This downward-sloping line will eventually cross the upward-sloping line for each metal oxide. Above the temperature at which the lines cross, carbon can reduce that metal oxide to the metal.

- Above 1000 K, carbon can reduce iron(II) oxide, FeO, to iron.

Iron manufacture using a blast furnace is especially successful because the line for conversion of carbon monoxide to carbon dioxide lies *just below* that for iron up until 1000 K, so conversion of CO into CO_2 is always more exergonic (Taking the Next Step 9.1) than the conversion of Fe into FeO at every temperature below 1000 K.

- Below 1000 K, carbon monoxide can reduce iron(II) oxide to iron.

Figure 9.7 shows that the downward-sloping C/CO line does not cross the upward-sloping lines for the oxides of titanium, aluminium, or calcium until above 1600 K, so we need to manufacture these elements by a different technique (electrolysis: Section 8.7 describes the Downs cell used for the manufacture of sodium).

KINETICS

CHAPTER 10

10.1 THE RATE OF REACTION

- For particles to react they must **collide** and **the collision must be energetic enough to overcome the activation energy**.

- The **activation energy** for a reaction, denoted by E_a in Figure 10.1, is the minimum energy necessary for a collision to lead to a successful reaction.

- The **rate of reaction** is the rate of change of concentration per unit time. We usually measure the rate in mol dm^{-3} s^{-1}. It is the gradient of the concentration–time graph at a particular time. A Deeper Look 10.1 explains how to define a unique rate of reaction that does not depend on which reactant (or product) is considered.

A DEEPER LOOK 10.1

What is the formal definition of rate?

To make sure that every reaction has a unique rate, we need to consider whether the particular substance whose concentration we are measuring is a reactant or a product (and to take into consideration the mole ratios). So for the ammonia synthesis (Section 6.3), one N_2 molecule reacts with three H_2 molecules to form two NH_3 molecules, so the unique rate is:

$$rate = -d[N_2]/dt = -\frac{1}{3}d[H_2]/dt = \frac{1}{2}d[NH_3]/dt$$

We need minus signs for both reactants as their concentrations fall over time.

10.2 FACTORS THAT AFFECT THE RATE OF REACTION

- The **rate of reaction depends on the concentration** of the reactants. (The equivalent for gases is their pressure.)
  - The higher the concentration, the more particles there are per unit volume. Hence, simplistically, the number of collisions is higher and the rate is faster. However, we need to note the critical role of the order of reaction (Section 10.3).

- The **rate of reaction depends on the temperature**.

In many cases, the particles must also collide in a particular orientation. See Section 10.5 on the Arrhenius equation.

(a)

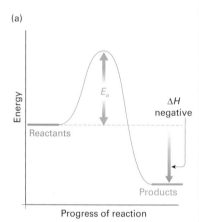

(b)

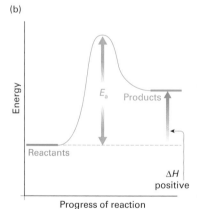

FIGURE 10.1 Reaction profiles for (a) exothermic and (b) endothermic reactions. E_a is the activation energy barrier that reactants must overcome before they can change into products. ΔH indicates the overall enthalpy change for the reaction. The point corresponding to the energy maximum represents the **transition state**.

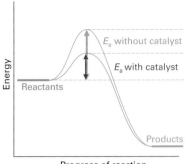

FIGURE 10.2 The reaction profiles for an uncatalysed reaction and the same reaction with a suitable catalyst present.

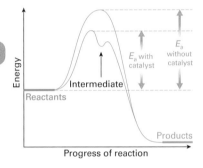

FIGURE 10.3 It is possible that the catalysed reaction involves an **intermediate**, unlike the situation shown in Figure 10.2.

To catalyse ester hydrolysis (Section 23.4), adding sodium hydroxide introduces a better nucleophile (hydroxide ion), so Figure 10.2 is appropriate. Adding hydrochloric acid, however, creates an intermediate (the protonated ester), so Figure 10.3 is appropriate.

- Increased temperature increases the number of collisions per second.
- However, the more important factor is that the number of particles that have the necessary activation energy increases significantly (see the explanation below using the Maxwell–Boltzmann distribution, Section 10.4).

• The **rate of reaction depends on the presence of a catalyst**.
  - A **catalyst** is a substance that increases the rate of a chemical reaction and is recoverable at the end unchanged chemically and in mass.
  - A catalyst provides an **alternative reaction route** with a **lower activation energy**, as shown in Figures 10.2 and 10.3.
  - A catalyst for the forward reaction also catalyses the backward reaction (Section 21.3), as can be seen from Figure 10.2. The catalyst reduces the activation energy of the backward reaction by the same amount as it reduces the activation energy of the forward reaction.
  - There are both **heterogeneous and homogeneous** catalysts (Section 14.3).

• The **rate of reaction depends on the state of subdivision** of the particles:
  - The higher the surface area, the faster the reaction: the increased surface area allows more collisions to take place. So powdered reactants produce a faster rate of reaction than lumps of the same substance.

10.3 THE EFFECT OF CONCENTRATION ON REACTION RATE: A MORE DETAILED VIEW

• The **order** with respect to a substance is the power to which its concentration is raised in the experimentally determined rate equation.
  - If the **rate equation** is $rate = k[A]^m[B]^n$
    ○ m is the order with respect to A and n is the order with respect to B.
    ○ The **overall order** is the sum of the individual orders.
    ○ k is the **rate constant** and its unit depends on the order.

For a zero-order reaction, $rate = k$, so k has the same unit as $rate$, mol dm^{-3} s^{-1}.
  For a first-order reaction, $rate = k[A]$; mol dm^{-3} cancels on both sides, so k has the unit of s^{-1}.

• One way of finding order requires doing several experiments, each measuring the *initial rate* of the reaction as a function of the *initial concentration* of one of the reactants.
  - Comparing two sets of data, for example using double the concentration:
    ○ if the rate of reaction **does not change** as the concentration of the reactant doubles, the reaction is **zero order** with respect to that reactant.

TABLE 10.1 Experimental data for the peroxodisulfate/iodide reaction.

Experiment no.	Initial concentrations/ mol dm^{-3}		Initial rate of reaction/mol dm^{-3} s^{-1}
	[S$_2$O$_8^{2-}$]	[I$^-$]	
1	0.038	0.030	7.0 × 10^{-6}
2	0.076	0.030	14.0 × 10^{-6}
3	0.076	0.060	28.0 × 10^{-6}

- ○ if the rate of reaction **doubles** as the concentration of the reactant doubles, the reaction is **first order** with respect to that reactant.
- ○ if the rate of reaction **quadruples** as the concentration of the reactant doubles, the reaction is **second order** with respect to that reactant.

Table 10.1 shows experimental data for the reaction between peroxodisulfate ion S$_2$O$_8^{2-}$ and iodide ion I$^-$. Comparing experiments 1 and 2, the I$^-$ ion concentration is constant and the S$_2$O$_8^{2-}$ ion concentration doubles, which causes the rate to double: the order with respect to S$_2$O$_8^{2-}$ is one. (Comparing experiments 2 and 3, the order with respect to I$^-$ is also one.)

- Another way of finding order requires only a *single* experiment, but we need to solve the rate equation to work out what graph to draw to find the order of the reaction.

- A **zero-order** reaction has a constant gradient for the concentration–time graph: the rate is constant because it does not depend on concentration. Plotting a graph of the **concentration c against time t** therefore gives a straight line: the equation of the line is

$$c = c_0 - kt$$

where c_0 is the initial concentration (at time 0).

  - Figure 10.4 shows the graph of iodine concentration against time for the reaction between propanone and iodine ($CH_3COCH_3 + I_2 \rightarrow CH_3COCH_2I + HI$): the gradient is $-k$.
  - The straight line tells us that this is a zero-order reaction.

  Finding the order of reaction gives vital information about its **mechanism**: exactly what happens at every stage of the reaction. So the fact that the order with respect to iodine for the iodination of propanone is zero means that the **rate-limiting step** (**rate-determining step**) cannot involve iodine: see A Deeper Look 22.4, where we discuss the actual mechanism for this reaction in detail.

- A **first-order** reaction has a concentration–time graph that shows an exponential decay, as shown in Figure 10.5. Taking the Next Step 10.1 explains how to solve a first-order rate equation.

The graph shows that the concentration falls exponentially. The half life $t_{1/2}$, discussed below, is constant for a first-order reaction.

(a) Initial concentrations

iodine	0.01 mol dm^{-3}
propanone	0.25 mol dm^{-3}
sulfuric acid	0.25 mol dm^{-3}

At $t = 0$, $[I_2] = 0.01$ mol dm^{-3}

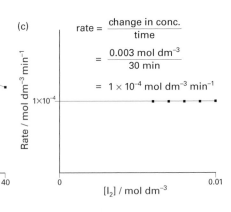

(c)

$$rate = \frac{\text{change in conc.}}{\text{time}}$$

$$= \frac{0.003 \text{ mol dm}^{-3}}{30 \text{ min}}$$

$$= 1 \times 10^{-4} \text{ mol dm}^{-3} \text{ min}^{-1}$$

FIGURE 10.4 The experiment to determine the order of reaction with respect to iodine for the reaction between iodine and propanone (acetone). (a) Initial concentrations of the reaction mixture. (b) The results plotted as a graph (iodine concentration vs time). (c) The rate is constant.

10

FIGURE 10.5 A graph of the concentration of N_2O_5 against time for the reaction:

$2N_2O_5(g) \rightarrow 4NO_2(g) + O_2(g)$ at 70 °C.

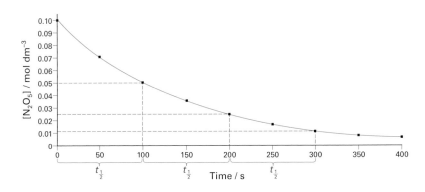

TAKING THE NEXT STEP 10.1

How can we solve a first-order rate equation?

We can express the rate of reaction mathematically as $rate = -dc/dt$, where c is the concentration of a reactant (see A Deeper Look 10.1). For a first-order reaction, $rate = kc$

Substituting into $rate = -dc/dt$ and multiplying by −1 gives

$$dc/dt = -kc$$

The solution of this equation is

$$c = c_0 e^{-kt}$$

where c_0 is the initial concentration (at time 0). To prove this, differentiate with respect to t, remembering that c_0 and k are constant:

e is the number 2.718... that is the base of natural logarithms (ln). If $y = e^{ax}$ then $dy/dx = a\,e^{ax}$. $\ln xy = \ln x + \ln y$; $\ln e^x = x$.

$$dc/dt = c_0 \times -ke^{-kt} = -k\,c_0 e^{-kt} = -kc$$

Taking natural logarithms of both sides of the equation $c = c_0 \, e^{-kt}$ gives

$\ln c = \ln c_0 - kt$

Therefore, plotting the graph of the **natural logarithm of the concentration ($\ln c$) against time** gives a straight line, with a gradient equal to $-k$.

Figure 10.6 shows the graph of the natural logarithm of the concentration of dinitrogen pentoxide, N_2O_5, against time for its decomposition into nitrogen dioxide and oxygen, using the data from Figure 10.5.

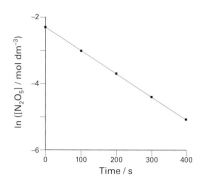

- The **half-life** is the time taken for the concentration to fall to half its original value. Figure 10.5 shows that the **half-life is constant for a first-order reaction**. Taking the Next Step 10.2 explores the link between the half-life and the rate constant. A Deeper Look 10.2 shows that the half-life *does* depend on concentration for a second-order reaction.

FIGURE 10.6 The reaction $2N_2O_5(g)$ $\rightarrow 4NO_2(g) + O_2(g)$ is first order with respect to the reactant, as the *natural log* graph is a straight line.

Radioactive decay is a first-order process. The metastable radionuclide (Section 1.1) technetium-99m used for bone scans has the useful half-life of 6.0 hours.

TAKING THE NEXT STEP 10.2

How is the half-life related to the rate constant?

For a *first*-order reaction, we can relate the half-life to the rate constant using the equation

$\ln c = \ln c_0 - kt$ (derived in Taking the Next Step 10.1) by making kt the subject and then combining the natural logarithms:

$kt = \ln c_0 - \ln c = \ln(c_0/c)$

Maths reminder: $\ln(x/y) = \ln x - \ln y$

The half-life $t_{1/2}$ is the time taken for the concentration to fall to half its initial value, so when

$t = t_{1/2}$, $c = \frac{1}{2}c_0$ and hence $kt_{1/2} = \ln 2$ or

$t_{1/2} = (\ln 2)/k$

Remember that this equation is only true for a **first**-order reaction.

A DEEPER LOOK 10.2

What about second-order reactions?

For a second-order reaction, the rate equation is

$dc/dt = -kc^2$

The solution of this equation is

$\dfrac{1}{c} = \dfrac{1}{c_0} + kt$ (1)

Plotting a graph of the **reciprocal of concentration ($1/c$) against time** gives a straight line.

10

Substituting the definition of the half-life (Taking the Next Step 10.2) into (1):

$$\frac{2}{c_0} = \frac{1}{c_0} + kt_{1/2}$$

hence the equation for the half-life for a second-order reaction is

$$t_{1/2} = \frac{1}{kc_0}$$

which shows that the half-life is inversely proportional to the initial concentration for a second-order reaction. Therefore, the half-life doubles as the concentration halves, as shown in Figure 10.7. Remember that the half-life for a *first*-order reaction is independent of concentration (see Taking the Next Step 10.2).

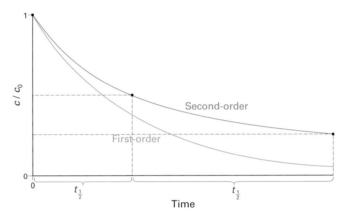

FIGURE 10.7 The half-life of a second-order reaction doubles as the concentration halves.

10.4 THE EFFECT OF TEMPERATURE ON THE RATE OF REACTION: A MORE DETAILED VIEW

At university you may also come across a similar graph for the distribution of the *speeds* of the molecules.

- The **Maxwell–Boltzmann distribution**, as shown in Figure 10.8, plots the number of molecules in a gas with a given energy against energy. We can see that:
  - no molecules have zero energy;
  - the number of molecules with very high energy is small: the graph is asymptotic at high energy;
  - the area under the graph is equal to the total number of molecules.
- The area under each distribution curve represents the total number of molecules.

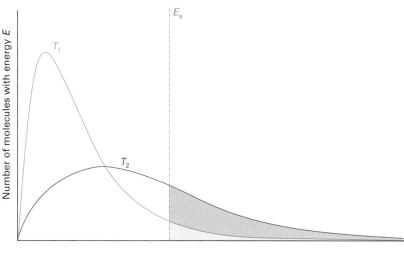

FIGURE 10.8 The distribution of molecular energies at two temperatures (T_2 is significantly higher than T_1).

- There is a wider range of molecular energies at the higher temperature (T_2). The peak of the curve lies at higher energies at higher temperatures; the average energy also increases at higher temperatures.

- The vertical line labelled E_a represents the activation energy. The number of molecules that have at least this energy is much greater at the higher temperature (shown by the shading).

- Only the molecules with energy above E_a will react. So you might ask why the reaction does not stop when these molecules are depleted. You actually need to think of this graph as a dynamic process. While the molecules with enough energy react, the others will keep on bumping into each other, transferring energy to one another, and therefore replenishing the number of molecules above E_a.

- At higher temperatures, as seen in Figure 10.8,
  - the average energy increases;
  - the maximum occurs at a higher energy;
  - the maximum is lower, as the area must remain the same (as the number of molecules is fixed);
  - ***most significantly*** more molecules have high energy.

- As argued above (Section 10.2), the most important factor explaining the significant effect of temperature on reaction rate is that **increasing the temperature causes a large increase in the number of molecules with the necessary activation energy**.

- A catalyst provides an alternative reaction route with a lower activation energy, so more molecules have the necessary activation energy, as shown in Figure 10.9, and the reaction happens faster.

10

FIGURE 10.9 A catalyst provides an alternative reaction route with a lower activation energy, so more particles will have the necessary activation energy (as shown by the shading).

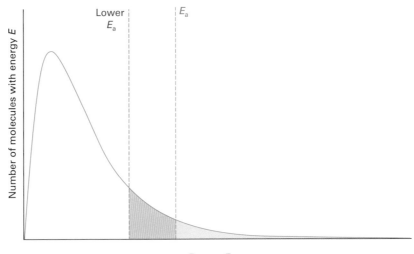

10.5 THE ARRHENIUS EQUATION

- We can measure the activation energy E_a using the **Arrhenius equation**, which shows how the rate constant k depends on the temperature T:

  - $k = A\, e^{-E_a/RT}$

    ○ where A is the **Arrhenius factor**, R is the gas constant, and T is the thermodynamic temperature.

    ○ As the exponential term has no units, the units of A are the same as the units of k and therefore depend on the order of reaction (Section 10.3).

    > The Arrhenius factor takes into account the collision rate, together with the possibility that the orientation of the collision is important. The exponential term represents the fraction of molecules with an energy greater than the activation energy.

Other names for A that you may come across are the **frequency factor** or the **pre-exponential factor**.

Taking natural logarithms of both sides,

  - $\ln k = \ln A - E_a/RT$

    ○ This has the form of a straight line if you **plot ln k against 1/T**. The gradient of the line equals $-E_a/R$ (and the intercept is $\ln A$).

    ○ Hence we must measure the rate constant k as a function of temperature T and then draw an **Arrhenius plot** of ln k against 1/T. Figure 10.10 shows one example.

At university, you will meet some reactions whose temperature dependence is not adequately explained using the Arrhenius equation.

- Rates of reaction often vary greatly with temperature. Using the Arrhenius equation, with an activation energy of 50 kJ mol^{-1}, rate roughly doubles from 15 °C to 25 °C.

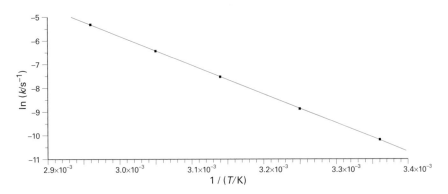

FIGURE 10.10 An Arrhenius plot for the decomposition of dinitrogen pentoxide at different temperatures (shown in Table 10.2). The gradient of the line is $-E_a/R$: for this reaction, E_a is 105 kJ mol^{-1}.

TABLE 10.2

$T/°C$	k/s^{-1}	$\ln(k/s^{-1})$	$1/(T/K)$
25	3.4×10^{-5}	-10.29	3.36×10^{-3}
35	1.4×10^{-4}	-8.87	3.25×10^{-3}
45	5.0×10^{-4}	-7.60	3.14×10^{-3}
55	1.5×10^{-3}	-6.50	3.05×10^{-3}
65	4.9×10^{-3}	-5.32	2.96×10^{-3}

CHAPTER

11

TRENDS ACROSS THE PERIODIC TABLE

11.1 PHYSICAL TRENDS ACROSS THE PERIODIC TABLE

- The nature of the elements changes significantly across the periodic table.
  - The most commonly studied periods are Period 2 (Li to Ne) and especially Period 3 (Na to Ar).
- On the **left** of the table the elements are **metals**, whereas on the **right** of the table the elements are **non-metals**. The stepped line drawn rather like a staircase in Figure 11.1 roughly marks the division between the metals and non-metals. Taking the Next Step 11.1 explains that a more careful approach would acknowledge the existence of the **metalloids**.
  - In Period 3 the **electrical conductivity** is high for the metallic elements sodium, magnesium, and aluminium. Silicon is a semiconductor. The other elements are all insulators, having *very* low conductivities. Across any period, the elements change from conductors to insulators.

Period	Group 1	2											13	14	15	16	17	18
1	H 1																	He 2
2	Li 3	Be 4											B 5	C 6	N 7	O 8	F 9	Ne 10
3	Na 11	Mg 12											Al 13	Si 14	P 15	S 16	Cl 17	Ar 18
4	K 19	Ca 20	Sc 21	Ti 22	V 23	Cr 24	Mn 25	Fe 26	Co 27	Ni 28	Cu 29	Zn 30	Ga 31	Ge 32	As 33	Se 34	Br 35	Kr 36
5	Rb 37	Sr 38	Y 39	Zr 40	Nb 41	Mo 42	Tc 43	Ru 44	Rh 45	Pd 46	Ag 47	Cd 48	In 49	Sn 50	Sb 51	Te 52	I 53	Xe 54
6	Cs 55	Ba 56	La 57	Hf 72	Ta 73	W 74	Re 75	Os 76	Ir 77	Pt 78	Au 79	Hg 80	Tl 81	Pb 82	Bi 83	Po 84	At 85	Rn 86
7	Fr 87	Ra 88	Ac 89	Rf 104	Db 105	Sg 106	Bh 107	Hs 108	Mt 109	Ds 110	Rg 111	Cn 112	Nh 113	Fl 114	Mc 115	Lv 116	Ts 117	Og 118

Ce 58	Pr 59	Nd 60	Pm 61	Sm 62	Eu 63	Gd 64	Tb 65	Dy 66	Ho 67	Er 68	Tm 69	Yb 70	Lu 71
Th 90	Pa 91	U 92	Np 93	Pu 94	Am 95	Cm 96	Bk 97	Cf 98	Es 99	Fm 100	Md 101	No 102	Lr 103

FIGURE 11.1 The standard modern form of the periodic table. The deeper green colour identifies the metalloids. The **s block** is in yellow, the **d block** in blue, the **p block** in green, and the **f block** in red.

Which elements are metalloids?

Five of the chemical elements are intermediate in nature between metals and non-metals. Silicon (Si) and germanium (Ge), for example, are semiconductors rather than conductors or insulators. Together with arsenic (As), antimony (Sb), and tellurium (Te), they are called **metalloids**. Any element to the left or below the metalloids is a metal. Any element to the right or above the metalloids is a non-metal.

11.2 MELTING POINTS ACROSS PERIODS 3 AND 2

- The melting point of a metallic element depends on the strength of the metallic bonding (Section 4.4) it experiences, and hence the **attraction between the metal ions and the delocalized sea of electrons**.
  - **Aluminium has the highest melting point** because its ion Al^{3+} has the highest charge density (Section 2.7) and aluminium contributes the largest number of electrons (3) into the delocalized sea.
  - **The melting point gradually rises** from Na to Mg to Al (although not smoothly), as shown in Figure 11.2.
- The next element after aluminium is silicon. Silicon is a **giant covalent solid** (Section 4.4) and as such its very strong covalent bonds need to be weakened to allow melting.
  - **Si therefore has the highest melting point** in Period 3, as seen in Figure 11.2.
- Figure 11.2 shows that the next four elements (phosphorus, sulfur, chlorine, and argon) all have **very low melting points**, because the forces of attraction between the molecules are only **dispersion forces**.

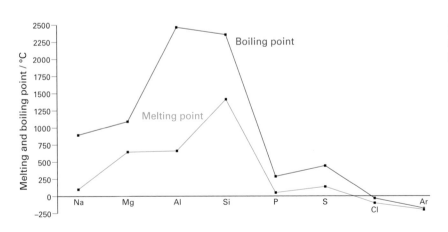

FIGURE 11.2 Trends in the melting and boiling points of the Period 3 elements.

FIGURE 11.3 Trends in the melting and boiling points of the Period 2 elements are very similar to the Period 3 trends, except that the values for nitrogen, oxygen, and fluorine are almost identical as they are all diatomic (N_2, O_2, F_2).

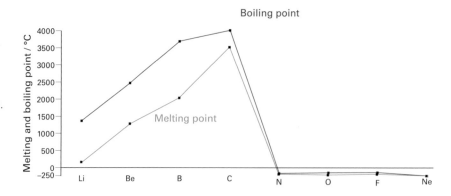

- We can explain the variation in melting point across these four elements by knowing the molecular formula of the element: **P_4, S_8, Cl_2,** and **Ar**.

- As dispersion forces increase as the number of electrons in a molecule increases (Section 4.2), sulfur (S_8) has the highest melting point, followed by phosphorus (P_4) then chlorine (Cl_2) then argon (Ar).

- Figure 11.3 shows that the trend in Period 2 is very similar to the trend in Period 3, except that the **melting points for nitrogen, oxygen, and fluorine are very similar** as they are all diatomic molecules.

11.3 ATOMIC RADII ACROSS PERIOD 3

Going across a period, **the atoms get smaller**, as shown in Figure 11.4.

- Each successive element has **one more proton and one more electron**. The **extra electron does not shield the other electrons in the same shell effectively**. Taking the Next Step 11.2 explains the concept of shielding in more detail.

FIGURE 11.4 Trends in atomic radius for Periods 2 and 3.

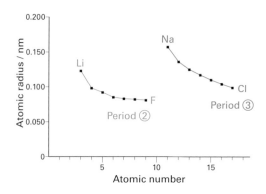

TAKING THE NEXT STEP 11.2

How does shielding change across Period 3?

As we move across a period, each successive element has one more proton and one more electron. The extra electron does not shield the other electrons in the same shell effectively. To understand why this is the case, we need to understand the concept of effective nuclear charge.

We define the **effective nuclear charge Z_{eff}** as the actual nuclear charge minus the shielding constant (which takes into account the shielding effect of all the other electrons). Going across a period the effective nuclear charge increases, hence there is a stronger attraction to the nucleus felt by the outermost electrons. As a result, the atom becomes smaller.

John Slater introduced a simple model for approximating the value of the effective nuclear charge. **Slater's rules** suggest that an electron in the same shell contributes only about 0.35 to the shielding constant, electrons in the penultimate shell contribute about 0.85, and electrons in shells even closer to the nucleus contribute 1 (hence shielding effectively perfectly). Slater's rules applied to Na would give a Z_{eff} value of $11 - (8 \times 0.85) - (2 \times 1) = 2.2$. The rules would give a Z_{eff} value for Si of $14 - (3 \times 0.35) - (8 \times 0.85) - (2 \times 1) = 4.15$ and a Z_{eff} value for Ar of $18 - (7 \times 0.35) - (8 \times 0.85) - (2 \times 1) = 6.75$. Notice how these approximate values are close to the more exact calculated values shown in Table 11.1.

TABLE 11.1 Calculated effective nuclear charges across Period 3.

Atom	Electron ionized	Z_{eff}
Na	3s	2.51
Mg	3s	3.31
Al	3p	4.07
Si	3p	4.29
P	3p	4.89
S	3p	5.48
Cl	3p	6.12
Ar	3p	6.76

11.4 IONIZATION ENERGIES ACROSS PERIOD 3

- Figure 11.5 shows how **sodium has a low ionization energy** (498 kJ mol^{-1}) compared with the preceding element (neon, 2081 kJ mol^{-1}), because the electron being ionized is in a **new shell** (shell 3 rather than shell 2), further away from the nucleus.

FIGURE 11.5 Plot of first ionization energy against atomic number for $Z =$ 11 (Na) to $Z = 18$ (Ar). The plot shows that there is a periodic variation in ionization energy across Period 3.

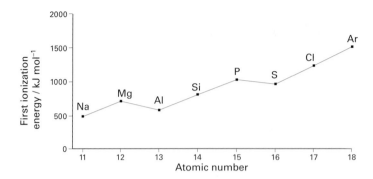

- Figure 11.5 shows that there is **another, less significant, drop in ionization energy for aluminium** compared with magnesium, because the electron being ionized is in a **new subshell** (3p rather than 3s), which is higher in energy.

- The rise from sodium to magnesium, the rise from aluminium to phosphorus, and the rise from sulfur to argon, Figure 11.5, all occur because the **number of protons increases by one** for each successive element and the **extra electron added as well is poor at shielding** (see Taking the Next Step 11.2).

- The **drop in ionization energy from phosphorus to sulfur** is much more difficult to explain.

  > You will probably have come across the explanation that this drop occurs because there is additional electron–electron repulsion between the two electrons in one orbital for sulfur. Such an explanation, while sounding convincing, is not in agreement with detailed calculations.

11.5 CHEMICAL TRENDS ACROSS THE PERIODIC TABLE

- The **oxidation state** (Section 8.2) of the elements changes significantly across the periodic table, as judged especially by their reaction with oxygen.

- In Period 3, the **oxidation number of the highest oxide gradually increases by one** from Ox(Na) = +1 in Na_2O to Ox(Cl) = +7 in Cl_2O_7.
  - It was this smooth trend which convinced Mendeleyev that his periodic table would be very valuable.

- **Phosphorus, sulfur, and chlorine form a variety of oxides**. See Taking the Next Step 11.3.
  - Phosphorus forms P_4O_6 and P_4O_{10}.
  - Sulfur forms SO_2 and SO_3.
  - Chlorine forms several oxides including Cl_2O, ClO_2, and Cl_2O_7.

TAKING THE NEXT STEP **11.3**

Naming the oxides

The way of naming the oxides depends on the nature of the element. For the metals (sodium, magnesium, and aluminium), the names are sodium oxide, magnesium oxide, and aluminium oxide. We do not need to include an oxidation state, as we need to do in the case of iron(III) oxide for example, because these elements only have one oxidation state.

For metalloids and non-metals, the name usually reflects the formula of the compound—either the molecular formula for a gaseous oxide or the empirical formula for a solid oxide. So silicon dioxide is the name for SiO_2, sulfur dioxide for SO_2, sulfur trioxide for SO_3, dichlorine oxide for Cl_2O, chlorine dioxide for ClO_2, and dichlorine heptoxide for Cl_2O_7.

An exception to this rule is that P_4O_6 and P_4O_{10}, which would consistently be called tetraphosphorus hexoxide and tetraphosphorus decoxide, have recommended names of phosphorus(III) oxide and phosphorus(V) oxide.

Phosphorus(V) oxide, an important dehydrating agent, was traditionally called phosphorus pentoxide as the solid has an empirical formula (Section 3.1) of P_2O_5.

11.6 REACTION WITH OXYGEN ACROSS PERIOD 3

- All the elements from sodium to sulfur burn in oxygen or air, forming an oxide.
- Sodium burns with a **yellow flame** (Section 1.3 discusses the strong yellow line in sodium's atomic emission spectrum).
- Most of them (magnesium, aluminium, silicon, and phosphorus) generate **white light** from the strongly exothermic reaction and produce a **white smoke** containing the element's oxide.
- Sulfur burns with a **blue flame** and produces a **pungent colourless gas**.

$$4Na(s) + O_2(g) \rightarrow 2Na_2O(s) \text{ (see A Deeper Look 11.1)}$$
$$2Mg(s) + O_2(g) \rightarrow 2MgO(s)$$
$$4Al(s) + 3O_2(g) \rightarrow 2Al_2O_3(s)$$
$$Si(s) + O_2(g) \rightarrow SiO_2(s)$$
$$P_4(s) + 5O_2(g) \rightarrow P_4O_{10}(s)$$
$$S(s) + O_2(g) \rightarrow SO_2(g)$$

Taking the Next Step 11.4 discusses the nature of the oxides across Period 3.

This is the usual equation for the combustion of sulfur. (In the manufacture of sulfuric acid, the temperature is such that the sulfur is liquid and consists of individual atoms.) Given that the solid contains S_8 molecules (Section 11.2), it would also be correct to write $S_8(s) + 8O_2(g) \rightarrow 8SO_2(g)$.

A DEEPER LOOK **11.1**

What else can Group 1 elements form with oxygen?

In addition to the oxide, sodium also forms some sodium peroxide, Na_2O_2, which contains the **peroxide ion O_2^{2-}**. In addition to the oxide and peroxide, potassium forms some potassium superoxide KO_2, which contains the **superoxide ion O_2^-** (Section 2.1).

Potassium superoxide is used for purifying the air in submarines by reacting with carbon dioxide to form oxygen:

$$4\,KO_2(s) + 2\,CO_2(g) \rightarrow 2\,K_2CO_3(s) + 3\,O_2(g)$$

TAKING THE NEXT STEP 11.4

How do the oxides change across Period 3?

Going across the table, we find significant changes in the **nature of the bonding in the oxides** and hence in the state in which we find the oxide, as seen in Figure 11.6.

- The metals sodium, magnesium, and aluminium form oxides (**Na$_2$O, MgO,** and **Al$_2$O$_3$**) which are predominantly **ionic** and so exist as **solids with high melting points**.

- The metalloid silicon forms silicon dioxide (**SiO$_2$**) which is predominantly covalent and adopts a **giant covalent** structure, shown in Figure 11.7, which is therefore also a **solid with a high melting point**.

- The oxides of the non-metals phosphorus and sulfur also have covalent bonding, but these oxides are **simple molecular**: **P$_4$O$_{10}$** is a **low melting-point solid** (see A Deeper Look 11.2). Sulfur dioxide **SO$_2$** is a **gas**.

Taking the Next Step 11.5 discusses the changes in pH of solutions containing the oxides.

Silicon dioxide, also called silica, is a major constituent of sand.

11

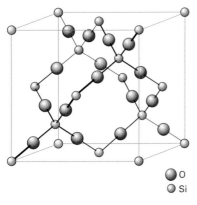

FIGURE 11.7 The silicon atoms in one form of silica adopt the diamond structure (Figure 4.19); there are oxygen atoms between each pair of silicon atoms.

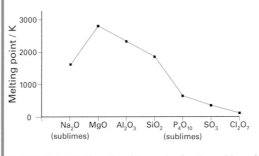

FIGURE 11.6 Plot of melting point for the oxides of the elements of Period 3. Note that, where an element has more than one oxide, we choose the oxide corresponding to the highest oxidation number.

A DEEPER LOOK 11.2

What are the structures of P$_4$O$_6$, P$_4$O$_{10}$, and SO$_3$?

The structure of P$_4$O$_6$ in the gas phase, shown in Figure 11.8(a), is based on a tetrahedron of phosphorus atoms with each phosphorus atom linked to three neighbours by oxygen atom bridges. (Note how the bottom part of the structure resembles the shape of cyclohexane.)

The structure of P$_4$O$_{10}$ in the gas phase, shown in Figure 11.8(b), is based on that of P$_4$O$_6$ but with each phosphorus atom having an additional double bond to an oxygen atom.

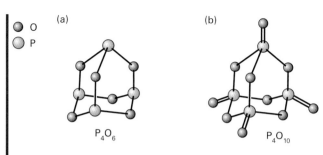

FIGURE 11.8 The structures of (a) P_4O_6 and (b) P_4O_{10}.

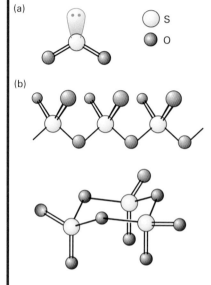

FIGURE 11.9 (a) The structure of an SO_2 molecule. (b) SO_3 molecules join together to make chains and rings.

Sulfur trioxide molecules join together in the solid state to form chains and rings (as shown in Figure 11.9). The ring structure explains why the melting point of SO_3 (see Figure 11.6) falls between those of P_4O_{10} and Cl_2O_7 as the solid is effectively S_3O_9.

TAKING THE NEXT STEP 11.5

How does the pH of the oxide's aqueous solution change?

There is a significant change in the pH of aqueous solutions containing the oxides from **alkaline for Na_2O to acidic for SO_3 and the chlorine oxides**. A Deeper Look 11.3 explains why this happens.

Na_2O reacts with water, as illustrated in Figure 11.10, to form the strong alkali (Section 7.2) sodium hydroxide. The pH of the solution is close to **14**:

• $Na_2O(s) + H_2O(l) \rightarrow 2NaOH(aq)$

MgO also reacts very slowly to form the *weak* base (Section 7.3) magnesium hydroxide, hence the pH of the solution is only **10**:

• $MgO(s) + H_2O(l) \rightarrow Mg(OH)_2(aq)$

Neither **Al_2O_3** nor **SiO_2** dissolves in water, so water remains at pH **7**.

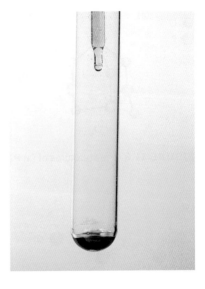

FIGURE 11.10 Solid sodium oxide $Na_2O(s)$ reacts with water exothermically. The product is aqueous sodium hydroxide $NaOH(aq)$, which turns the indicator blue.

You may have been told that sulfur dioxide is 'moderately soluble'. It is in fact the third most soluble gas.

P_4O_{10} reacts with water to form the weak acid (Section 7.3) phosphoric acid, and the pH of the solution is about **1**:

- $P_4O_{10}(s) + 6H_2O(l) \rightarrow 4H_3PO_4(aq)$

> You may have been told that phosphoric acid is a strong acid. In fact its pK_{a1} value is 2.15. The equilibrium measured by pK_{a1} (that is, the pK_a value for its first ionization) is $H_3PO_4(aq) + H_2O(l) \rightleftharpoons H_3O^+(aq) + H_2PO_4^-(aq)$.

SO_2 reacts with water to form the weak acid sulfurous acid, and the pH of the solution is about **1**:

- $SO_2(g) + H_2O(l) \rightarrow H_2SO_3(aq)$

High levels of sulfur dioxide in the atmosphere contribute to **acid rain**.

SO_3 reacts with water to form the strong acid (Section 7.2) sulfuric acid, so the pH of the solution is close to **0**:

- $SO_3(s) + H_2O(l) \rightarrow H_2SO_4(aq)$

There are a number of chlorine oxides, which all dissolve in water to form acidic solutions. For example, dichlorine oxide Cl_2O forms $HOCl$ (Section 13.6).

A DEEPER LOOK 11.3

How can we explain the trend from alkaline to acidic oxides?

Think of a part of a compound involving an element E and a hydroxide (E–O–H).

For an element of low electronegativity such as sodium, the Na–O bond is highly polarized towards the oxygen, and so it is reasonable to imagine this producing an Na^+ ion and a hydroxide ion OH^-. The hydroxide ion will make the solution alkaline.

The O–H bond is highly polarized towards the oxygen. An element of high electronegativity such as chlorine enhances this polarization by also withdrawing electron density towards itself. This therefore allows the loss of a proton H^+ from the O–H bond. The proton will make the solution acidic.

GROUP 2

12.1 INTRODUCTION TO GROUP 2

- We call this group of elements **Group 2** (which is the modern notation that uses the 1 to 18 numbering system, Section 1.5, and the one often used pre-university) or **Group II** (the traditional notation). An older name for the group was the alkaline earth metals.

- The elements in Group 2 all have a **full s subshell**:
  - Mg has the electronic structure $1s^2\ 2s^2\ 2p^6\ 3s^2$

- The elements in Group 2 are all **metals** with low electronegativities.

- We will consider beryllium and radium much less thoroughly than the others.
  - **Beryllium** is less studied because its compounds are toxic: the small beryllium ion can substitute for the magnesium ion in important biomolecules.
  - **Radium** is less studied because it is highly radioactive.

12.2 PHYSICAL PROPERTIES OF THE GROUP 2 ELEMENTS

- Figure 12.1 shows how the **atomic radius (and ionic radius) increases** down the group, as each successive element has one more shell of electrons, so the valence shell becomes further from the nucleus.

Group 2

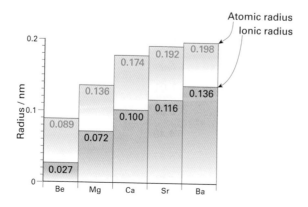

FIGURE 12.1 Each successive element in Group 2 has an extra shell of electrons. All ions are smaller than their parent atoms as the two electrons in the outer shell have been lost.

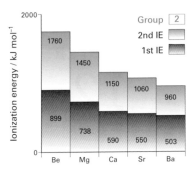

FIGURE 12.2 The first and second ionization energies of the Group 2 elements.

- The **ionization energy decreases** down the group, as the electron being ionized is in an orbital further from the nucleus; Figure 12.2 shows this trend.
  - The effect of shielding *down the group* needs *very* careful thought (see Taking the Next Step 12.1).

TAKING THE NEXT STEP 12.1

How does the shielding change down the group?

The effect of **shielding down the group** is usually taught incorrectly pre-university. To understand this concept, we need to focus on the effective nuclear charge Z_{eff} (Section 11.3).

Let us think very simply first. If the shielding down Group 2 was *completely perfect*, then Z_{eff} for each element would be 2 (minus a *small* shielding effect of the other electron in the valence shell). All the electrons in the inner shells would cancel all the protons except two. So this gives the simplest possible model of two electrons outside a completely shielded core. However, shielding cannot possibly be perfect, because, although an extra added proton is *certainly* in the nucleus, the extra added electron has a (small) probability of being at a great distance from the nucleus (Figure 1.7). Table 12.1 shows the calculated values for Z_{eff} for the valence electrons in Group 2.

The effective nuclear charge *increases* down the group. This appears to go against the observed trend in ionization energy, as a higher effective nuclear charge down the group would, on its own, make the ionization energy *larger* rather than smaller. However, the electron being ionized is in an orbital further from the nucleus, which would, on its own, make the ionization energy smaller. Thus the two factors (increasing Z_{eff} and increasing distance) operate in *opposite* directions. The greater distance from the nucleus has the bigger effect for Be to Ba, hence the ionization energy decreases from Be to Ba.

In your pre-university studies you will probably have stopped considering ionization energy data at Ba. The first ionization energy for Ra is 510 kJ mol⁻¹, a slightly ***larger value*** than for Ba (503 kJ mol⁻¹). It is at this point that Z_{eff} becomes the ***more*** important of the two factors. We can explain this by noting that there are now 32 extra electrons added between Ba and Ra, causing the effective nuclear charge to rise significantly.

Exactly the same effect happens in Group 1. Francium (380 kJ mol⁻¹) has a larger first ionization energy than caesium (376 kJ mol⁻¹).

TABLE 12.1 Calculated effective nuclear charges down Group 2.

Atom	Electron ionized	Z_{eff}
Be	2s	1.91
Mg	3s	3.31
Ca	4s	4.40
Sr	5s	6.07
Ba	6s	7.58

12

12.3 REACTIVITY WITH WATER AND ACIDS

- The Group 2 metals show **increasing reactivity with water** down the group. One major reason for this is that the ionization energy decreases down the group, which means that the outer electrons are lost more easily.

- **Magnesium reacts slowly with water**, as illustrated in Figure 12.3(a).
  - Reactivity increases significantly when **heated in steam**: the white solid magnesium oxide is formed.
    - $Mg(s) + H_2O(g) \rightarrow MgO(s) + H_2(g)$

- **Calcium reacts much faster with water than magnesium does**, as illustrated in Figure 12.3(b): we see bubbles of gas and a white cloudiness in the solution (the cloudiness is due to sparingly soluble calcium hydroxide):
  - $Ca(s) + 2H_2O(l) \rightarrow Ca(OH)_2(aq) + H_2(g)$

- Strontium and barium react with water faster than calcium does: the equations for the reactions are analogous to that for calcium.

- **Reactivity with acids is much faster than that with water**, producing a salt and hydrogen gas, e.g. $Mg(s) + 2HCl(aq) \rightarrow MgCl_2(aq) + H_2(g)$

12

12.4 THE SOLUBILITY OF THE HYDROXIDES

- The **solubility of the hydroxide increases** down the group: the variation is very significant, as shown in Table 12.2. Taking the Next Step 12.2 explains the variation.

- Magnesium hydroxide will precipitate easily when we add aqueous ammonia to an aqueous solution containing magnesium ions. Aqueous

FIGURE 12.3 (a) The rate of reaction of magnesium with cold water is slow. It does react more rapidly with steam. (b) Calcium reacts more vigorously with water. Calcium hydroxide has only limited solubility and gradually appears as a white cloudiness.

TABLE 12.2 The solubility of the Group 2 hydroxides.

Group 2 hydroxide	Solubility/g per100 cm³ H₂O
$Mg(OH)_2$	0.0009
$Ca(OH)_2$	0.18
$Sr(OH)_2$	0.41
$Ba(OH)_2$	5.6

The fact that magnesium hydroxide is very weakly basic explains its use as an antacid in indigestion tablets: the milky-white suspension is called '**milk of magnesia**'.

ammonia contains hydroxide ions (but because it is a weak base, Section 7.3, the concentration is typically less than 100th of that produced by NaOH). Once the mass of magnesium hydroxide exceeds 0.9 mg per 100 cm³ (see Table 12.2), no more will dissolve and the rest will precipitate. Calcium hydroxide is much less likely to precipitate as its solubility is much greater. A Deeper Look 12.1 explains how to understand this quantitatively.

- Calcium hydroxide is sufficiently soluble in water that the solution (lime water) is a weak alkali. This is particularly useful in the **lime water test** for carbon dioxide gas, a white milkiness in the solution confirming the presence of the gas:

  - $Ca(OH)_2(aq) + CO_2(g) \rightarrow CaCO_3(s) + H_2O(l)$

- **Barium hydroxide is a strong alkali** (Section 7.2) in water.

  - $Ba(OH)_2(aq) \rightarrow Ba^{2+}(aq) + 2OH^-(aq)$

TAKING THE NEXT STEP 12.2

Why does the solubility of the hydroxides change?

We can find the solution enthalpy (Section 5.4) of the hydroxides by adding together the endothermic lattice enthalpy of the solid and the exothermic hydration enthalpies of the metal ion and the two hydroxide ions. As the size of the metal ion increases down the group, the lattice enthalpy decreases (Table 5.2), from 3010 kJ mol⁻¹ for $Mg(OH)_2$ to 2340 kJ mol⁻¹ for $Ba(OH)_2$. The hydration enthalpy of the metal ion gets less exothermic, down from −1920 kJ mol⁻¹ for Mg^{2+} to −1310 kJ mol⁻¹ for Ba^{2+}. The lattice enthalpy falls more than the hydration enthalpy does (670 compared with 610), which makes the solution enthalpy of the hydroxides more negative down the group. Remember that there will, in addition, be a favourable *entropy change* (Section 9.1) accompanying solution because the ions locked in the solid can spread out into the solution.

A DEEPER LOOK 12.1

How can we explain the difference in ease of precipitation of magnesium and calcium hydroxides?

Magnesium hydroxide will precipitate easily when we add aqueous ammonia to an aqueous solution containing magnesium ions:

$Mg^{2+}(aq) + 2OH^-(aq) \rightarrow Mg(OH)_2(s)$

Calcium hydroxide is *much* less likely to precipitate. We can explain this difference quantitatively by considering the solubility product, which is the equilibrium constant for the reverse reaction:

$$Mg(OH)_2(s) \rightarrow Mg^{2+}(aq) + 2OH^-(aq)$$

We define this equilibrium constant in the same way as any other equilibrium constant (Section 6.1), except that we can omit the concentration of the *solid* because it is constant (being proportional to the density of the solid). So, for the equation above, the **solubility product** K_{sp} is

$$K_{sp} = [Mg^{2+}][OH^-]^2$$

Precipitation will occur when the product of the ionic concentrations exceeds the value of K_{sp}. The value of K_{sp} for $Mg(OH)_2$ is 1×10^{-11} mol^3 dm^{-9}, whereas that for $Ca(OH)_2$ is 6×10^{-6} mol^3 dm^{-9}. The hydroxide ion concentration in dilute aqueous ammonia is about 1×10^{-3} mol dm^{-3}, which means that $[OH^-]^2 = 1 \times 10^{-6}$ mol^2 dm^{-6}. So magnesium hydroxide will precipitate when the magnesium ion concentration exceeds 1×10^{-5} mol dm^{-3}. The calcium ion concentration, on the other hand, needs to exceed 6 mol dm^{-3} before precipitation occurs, which is very much less likely.

12.5 BARIUM SULFATE AND THE TEST FOR THE SULFATE ION

- The insolubility of barium sulfate provides a test for the sulfate ion.

- To test for a suspected sulfate, add **aqueous barium nitrate acidified with dilute nitric acid** to the solution: a **white precipitate** is a positive test.

  - $Ba^{2+}(aq) + SO_4^{2-}(aq) \rightarrow BaSO_4(s)$

  - The dilute nitric acid reacts with other ions, such as carbonate ion, that could interfere by precipitating.

Taking the Next Step 12.3 explains the trend in solubility of the sulfates down the group.

> Barite (a dense mineral comprising barium sulfate) is commonly used as a weighting agent for all types of drilling fluids. It can cause deposition of scale in pipes which is similar to the common lime scale found in hard water areas. Barium sulfate scale problems cost the oil and gas industry hundreds of millions of dollars per year in lost production.

TAKING THE NEXT STEP 12.3

How and why does the solubility of the sulfates change?

The **solubility of the sulfates decreases** down the group, as shown in Figure 12.5.

We can find the solution enthalpy (Section 5.4) of the sulfates by adding together the endothermic lattice enthalpy of the solid and the exothermic hydration enthalpies of the metal ion and the sulfate ion. Although the size

Barium sulfate is also used in medicine because as well as being insoluble it is opaque to X-rays, as illustrated in Figure 12.4.

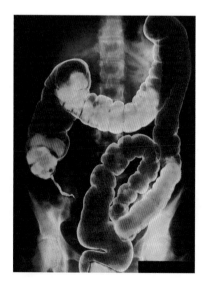

FIGURE 12.4 This radiograph shows the lower intestinal tract. The patient swallowed a suspension of barium sulfate about an hour before the X-ray was taken.

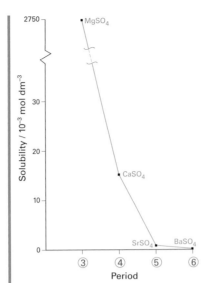

FIGURE 12.5 The solubility of Group 2 salts. The same decreasing trend as for sulfates is observed for other Group 2 salts with large, doubly charged negative ions such as carbonate CO_3^{2-}, ethanedioate $C_2O_4^{2-}$, and chromate(VI) CrO_4^{2-}.

of the metal ion increases down the group, the lattice enthalpy varies very little due to the very large size of the sulfate ion. (The values for $CaSO_4$, $SrSO_4$, and $BaSO_4$ are all within 3% of 2450 kJ mol⁻¹.) However, the hydration enthalpy of the metal ion becomes less exothermic, down from −1920 kJ mol⁻¹ for Mg^{2+} to −1310 kJ mol⁻¹ for Ba^{2+}. The lattice enthalpy falls much less than the hydration enthalpy does, which makes the solution enthalpy less negative down the group. Remember that there will, in addition, be a favourable *entropy change* (Section 9.1) accompanying solution because the ions locked in the solid can spread out into the solution.

THE HALOGENS

13.1 INTRODUCTION TO THE HALOGENS

- We call this group of elements **Group 17** (which is the modern notation using the 1 to 18 numbering system (Section 1.5), and is also typically used at university) or **Group VII** (the traditional name). The very common traditional name for the group is the **halogens**. We frequently use the symbol X for a general halogen.

- We will not consider **astatine**, because it is highly radioactive: its name comes from the Greek for 'unstable'.

13.2 PHYSICAL PROPERTIES OF THE HALOGENS

- The **melting and boiling points of the halogens increase** down the group, as seen in Figure 13.1. So fluorine and chlorine are gases, bromine is a liquid, and iodine is a solid (see Figure 4.8).

The only attractive forces between the X_2 molecules are dispersion forces (Section 4.2). These get larger as the number of electrons increases down the group.

- Figure 13.1 shows that the **atomic radius increases** down the group, as each successive element has one more shell of electrons.

  - Figure 13.2 shows that the **ionic radius also increases** down the group, as each successive element has one more shell of electrons.

- Figure 13.3 shows that the **electronegativity decreases** down the group, as the lower elements exert less attraction for a shared electron pair because they are larger.

Taking the Next Step 13.1 considers the trend in the ionization energy.

The name 'Group 7' is **incorrect** but is very common pre-university.

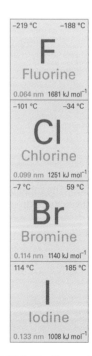

−219 °C	−188 °C
F	
Fluorine	
0.064 nm	1681 kJ mol⁻¹
−101 °C	−34 °C
Cl	
Chlorine	
0.099 nm	1251 kJ mol⁻¹
−7 °C	59 °C
Br	
Bromine	
0.114 nm	1140 kJ mol⁻¹
114 °C	185 °C
I	
Iodine	
0.133 nm	1008 kJ mol⁻¹

FIGURE 13.1 Top line: melting and boiling points. Bottom line: atomic radius and first ionization energy.

FIGURE 13.2 Halide ionic radii (to scale).

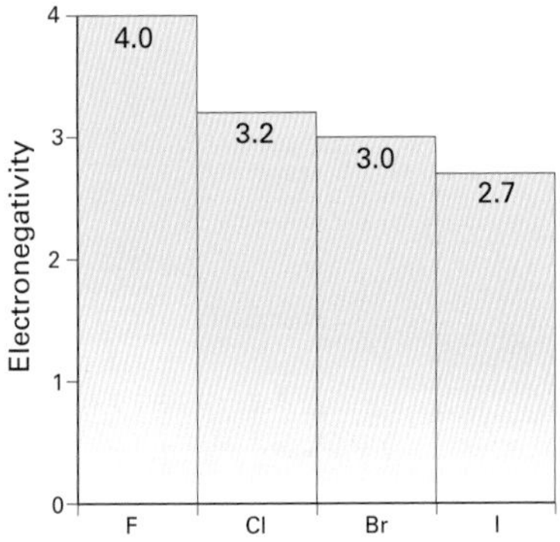

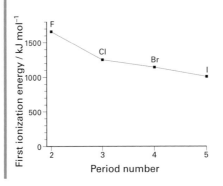

FIGURE 13.3 The variation in electronegativity down Group 17.

TAKING THE NEXT STEP 13.1

What is the trend in the ionization energy?

Figure 13.4 shows that the **ionization energy decreases** down the group, as the electron being ionized is further from the nucleus.

As explained in detail in the chapter on Group 2 (Section 12.2), the effective nuclear charge Z_{eff} **increases** down the group, so a pre-university argument that the shielding gets better down the group is incorrect. Table 13.1 shows the calculated values for Z_{eff}.

Thus the two factors (increasing Z_{eff} and increasing distance) operate in *opposite* directions, just as they do in Group 2 (Section 12.2). The greater distance from the nucleus has the bigger effect (as was the case for Be to Ba). Hence this justifies the simple explanation given above that the ionization energy decreases down the group because the electron being ionized is further from the nucleus.

FIGURE 13.4 The variation in ionization energy down Group 17.

TABLE 13.1 Calculated effective nuclear charges down Group 17.

Atom	Electron ionized	Z_{eff}
F	2p	5.10
Cl	3p	6.12
Br	4p	9.03
I	5p	11.61

13.3 OXIDIZING ABILITY OF THE HALOGENS

- Table 13.2 shows that the **oxidizing ability of the halogens X_2 decreases** down the group: the standard electrode potential, which measures oxidizing ability (Section 8.6), gets less positive down the group. Fluorine is exceptionally strongly oxidizing.

  - The reason for this trend involves processes such as the X–X bond enthalpy, the electron-gain enthalpy of X, and the hydration enthalpy of the halide ion.

  - One reason for the exceptionally high oxidizing ability of fluorine is the relatively weak F–F bond, as shown in Figure 13.5, which is

TABLE 13.2 Halogen standard electrode potentials.

Oxidized species	⇌	Reduced species	$E^{\ominus}$/V
$F_2(g) + 2e^-$	⇌	$2F^-(aq)$	+2.87
$Cl_2(g) + 2e^-$	⇌	$2Cl^-(aq)$	+1.36
$Br_2(l) + 2e^-$	⇌	$2Br^-(aq)$	+1.07
$I_2(s) + 2e^-$	⇌	$2I^-(aq)$	+0.54

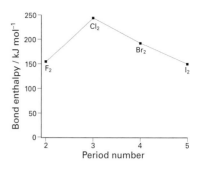

FIGURE 13.5 The variation in bond enthalpy down Group 17.

due to repulsion between the three lone pairs at each end of the bond getting close together. Another important factor is the large lattice enthalpies caused by the small fluoride ion or the large bond enthalpies caused by the small fluorine atom.

- The **consequences of the decreasing oxidizing ability** are:
  - **chlorine can displace bromine** from bromides, as illustrated in Figure 13.6, forming a yellow solution:
    - $Cl_2 + 2Br^- \rightarrow Br_2 + 2Cl^-$
  - **chlorine and bromine can displace iodine** from iodides, as illustrated for chlorine in Figure 13.6, forming a brown solution (or a black precipitate):
    - $Cl_2 + 2I^- \rightarrow I_2 + 2Cl^-$
    - $Br_2 + 2I^- \rightarrow I_2 + 2Br^-$

FIGURE 13.6 Halogen/halide displacement reactions: upper layer aqueous; lower layer tetrachloromethane. (a) Chlorine water added to fluoride: no reaction. (b) Chlorine oxidizes bromide: red bromine in the lower layer. (c) Chlorine oxidizes iodide: purple iodine in the lower layer.

  - iodine can displace none of the other halogens; fluorine can displace all the others but the reactions are very dangerously exothermic.

13.4 REDUCING ABILITY OF THE HALIDE IONS

- The **reducing ability of the halide ions X^- increases** down the group.
  - This follows from the decreasing oxidizing ability of the elements themselves, as seen in Table 13.2.

FIGURE 13.7 Reaction of potassium bromide and iodide with concentrated sulfuric acid. Note the yellow solid sulfur and the purple iodine vapour formed in the reaction with the iodide.

- Figure 13.7 illustrates this increasing strength of the halide ion as a reductant by showing the products of reaction of solid **potassium halides with concentrated sulfuric acid**.
  - **All four halides give steamy fumes of the hydrogen halide**, because the sulfuric acid protonates the halide ion (which is not a redox reaction):
    - $NaX + H_2SO_4 \rightarrow NaHSO_4 + HX$
  - **Chlorides produce no further reaction.**
  - **Bromides can reduce sulfuric acid to the colourless gas sulfur dioxide** (in which sulfur has oxidation number +4). Bromide ion is oxidized to brown bromine gas:
    - $H_2SO_4 + 2H^+ + 2Br^- \rightarrow SO_2 + 2H_2O + Br_2$
  - **Iodides can reduce sulfuric acid to the colourless gas sulfur dioxide or even further to produce yellow solid S or the gas hydrogen sulfide H$_2$S** (which smells of bad eggs and in which sulfur has oxidation number –2). Iodide ion is oxidized to purple iodine gas.
    - $H_2SO_4 + 8H^+ + 8I^- \rightarrow H_2S + 4H_2O + 4I_2$

13.5 THE TEST FOR AQUEOUS HALIDE IONS

- Figure 13.8 illustrates the test for aqueous halide ions: add **aqueous silver nitrate acidified with dilute nitric acid** (the nitric acid reacts with any other ions, such as carbonate ion, that would also produce precipitates).
  - **Fluoride ion does not form a precipitate** as silver fluoride is soluble.
  - **Chloride ion** forms a **white precipitate** of silver chloride:
    - $Ag^+(aq) + Cl^-(aq) \rightarrow AgCl(s)$

FIGURE 13.8 Tubes 1, 3, 5: Precipitates of silver chloride (white), silver bromide (cream), and silver iodide (yellow). The solids are light-sensitive and darken over time. Tubes 2, 4, 6: The effect on the precipitates of adding concentrated aqueous ammonia.

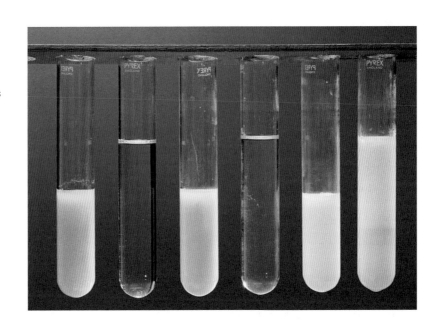

- **Bromide ion** forms a **cream precipitate** of silver bromide:
  - $Ag^+(aq) + Br^-(aq) \rightarrow AgBr(s)$
- **Iodide ion** forms a **yellow precipitate** of silver iodide:
  - $Ag^+(aq) + I^-(aq) \rightarrow AgI(s)$
- To make the test definitive, we then need to add **aqueous ammonia** (see Figure 13.8):
  - **Silver chloride** reacts with ('**redissolves** in') *dilute* or concentrated aqueous ammonia to form a colourless solution. See Taking the Next Step 13.2.
  - **Silver bromide** only reacts with ('**redissolves** in') *concentrated* aqueous ammonia to form a colourless solution.
  - **Silver iodide does not react** with aqueous ammonia.
  - The **complex ion** formed on reaction is $[Ag(NH_3)_2]^+$, Figure 15.8.

TAKING THE NEXT STEP 13.2

Why does silver chloride dissolve better than silver iodide?

Solubility involves a balance between lattice enthalpy and hydration enthalpy, Section 5.4. Both of these enthalpy changes become smaller in magnitude as the anion gets bigger from fluoride to iodide, as shown in Figure 13.2. Quantitatively, we can find the solution enthalpy by adding together the endothermic lattice enthalpy of the solid and the exothermic hydration enthalpies of the silver and halide ions, see Table 13.3. The

13

TABLE 13.3

Adding the enthalpies

Silver chloride	/kJ mol^{-1}
$\Delta_{lat}H^{\ominus}$	+915
$\Delta_{hyd}H^{\ominus}$ (Ag^+)	−472
$\Delta_{hyd}H^{\ominus}$ (Cl^-)	−377
$\Delta_{sol}H^{\ominus}$ (AgCl)	+66
Silver iodide	/kJ mol^{-1}
$\Delta_{lat}H^{\ominus}$	+889
$\Delta_{hyd}H^{\ominus}$ (Ag^+)	−472
$\Delta_{hyd}H^{\ominus}$ (I^-)	−305
$\Delta_{sol}H^{\ominus}$ (AgI)	+112
Silver fluoride	/kJ mol^{-1}
$\Delta_{lat}H^{\ominus}$	+967
$\Delta_{hyd}H^{\ominus}$ (Ag^+)	−472
$\Delta_{hyd}H^{\ominus}$ (F^-)	−515
$\Delta_{sol}H^{\ominus}$ (AgF)	−20

solution enthalpy for silver iodide is more endothermic than the solution enthalpy for silver chloride. Note that the solution enthalpy for silver fluoride is *exothermic*, so silver fluoride is soluble (remembering that the entropy change is always positive as the ions spread out into the solution).

The exothermic formation of the complex ion $[Ag(NH_3)_2]^+$ (together with the positive entropy change on solution) is sufficient to take AgCl into solution. Because a *complexation reaction* (Section 7.6) is involved when we add ammonia, the description that the precipitate 'redissolves' is inaccurate and should be avoided.

13.6 CHLORINE IN AQUEOUS (AND ALKALINE) SOLUTION

This is a **disproportionation** reaction (chlorine goes both down and up in oxidation number, Section 8.2).

• An equilibrium is set up when **chlorine 'dissolves' in water**: it **reacts** to form an aqueous solution containing both hydrochloric acid (in which chlorine has oxidation number −1) and hypochlorous acid (in which chlorine has oxidation number +1, see A Deeper Look 13.1).

– $Cl_2(g) + H_2O(l) \rightleftharpoons HCl(aq) + HOCl(aq)$

– The hypochlorous acid formed is a bleach so **chlorine bleaches moist litmus paper**.

• Because the aqueous mixture is acidic, we can significantly increase the extent of reaction by adding **sodium hydroxide**, which being alkaline reacts with the acids, causing the equilibrium to shift to the right. This forms a solution containing sodium chloride and sodium hypochlorite (the ionic equation emphasizes the fact that the sodium ion is a spectator ion):

– $Cl_2(g) + 2NaOH(aq) \rightarrow NaCl(aq) + NaClO(aq) + H_2O(l)$

– $Cl_2(g) + 2OH^-(aq) \rightarrow Cl^-(aq) + ClO^-(aq) + H_2O(l)$

• Chlorine compounds such as sodium hypochlorite are used in bleaches and to treat water in swimming pools.

A DEEPER LOOK 13.1

How do we name the halogen oxoacids and their salts?

Two different nomenclature systems are commonly used for naming the oxoacids of the halogens. (In the formulae for the acids given below, we emphasize the fact that hydrogen is bonded to *oxygen* rather than chlorine in each oxoacid.) We give the traditional name first and then the alternative name, which emphasizes the oxidation state of the chlorine:

HOCl is called hypochlorous acid or chloric(I) acid
HOClO (also written as $HClO_2$) is called chlorous acid or chloric(III) acid
HOClO$_2$ (also written as $HClO_3$) is called chloric acid or chloric(V) acid

$HOClO_3$ (also written as $HClO_4$) is called perchloric acid or chloric(VII) acid

Perchloric acid is a strong acid (Section 7.2) because of the extensive delocalization (Section 2.3) in the perchlorate ion ClO_4^-. Perchloric acid is also very dangerously exothermic when it reacts with organic matter. We name the salts formed from these acids in a similar way:

$NaClO$ is called sodium hypochlorite or sodium chlorate(I),
$NaClO_2$ is called sodium chlorite or sodium chlorate(III),
$NaClO_3$ is called sodium chlorate or sodium chlorate(V),
$NaClO_4$ is called sodium perchlorate or sodium chlorate(VII).

Once the oxoacid has been ionized, the anion formed is fully delocalized (Section 2.3). So sodium perchlorate in solution is present as $Na^+(aq)$ and $ClO_4^-(aq)$.

13

TRANSITION METALS 1

14.1 INTRODUCTION TO THE TRANSITION METALS

At university, it is very common to study the first row of the d-block elements (Sc to Zn) all together and hence avoid distinguishing between a transition element and a d-block element.

- A **transition metal (transition element)** is a d-block element with at least one stable **ion** that has a ***partially-filled d subshell***.

- The **first row of the transition metals** therefore consists of the elements **Ti to Cu**. The next two elements below each of Ti to Cu are also transition metals.

  - Scandium is not a transition metal, because it has only one oxidation state and the Sc^{3+} ion formed has *no* electrons in the d subshell.

  - Zinc is not a transition metal, because it has only one oxidation state and the Zn^{2+} ion formed has ten d electrons which *fill* the d subshell.

- Characteristic **transition metal properties** include

  - variable oxidation states

  - catalytic behaviour

  - complex formation

  - formation of coloured ions.

    One other characteristic is that first-row transition metal ions are frequently paramagnetic (Section 2.1). We can explain why (in Chapter 15) using the same argument that explains the colour of transition metal ions.

14.2 TRANSITION METALS SHOW VARIABLE OXIDATION STATES

Table 14.1 shows the main oxidation numbers shown in transition metal compounds.

- Looking at Table 14.1, we see that **the most common oxidation numbers are +2 and +3**: the classic example is iron forming iron(II) ions Fe^{2+} and iron(III) ions Fe^{3+}. The relative stability of these two states varies considerably across the first row; see Taking the Next Step 14.1.

TABLE 14.1 The common oxidation numbers (**bold** type indicates the most common) seen in the transition metals Ti to Cu.

oxidation number	Ti	V	Cr	Mn	Fe	Co	Ni	Cu
+7				+**7**				
+6			+**6**	+6	+6			
+5		+**5**						
+4	+**4**	+4	+4	+4		+4		
+3	+3	+3	+**3**	+3	+**3**	+3	+3	
+2	+2	+2	+2	+**2**	+**2**	+**2**	+**2**	+**2**
+1								+1
Element	**Ti**	**V**	**Cr**	**Mn**	**Fe**	**Co**	**Ni**	**Cu**
Electronic structure	[Ar] $3d^2 4s^2$	[Ar] $3d^3 4s^2$	[Ar] $3d^5 4s^1$	[Ar] $3d^5 4s^2$	[Ar] $3d^6 4s^2$	[Ar] $3d^7 4s^2$	[Ar] $3d^8 4s^2$	[Ar] $3d^{10} 4s^1$

TAKING THE NEXT STEP 14.1

How does the relative stability of the +2 and +3 ions change?

Figure 14.1 shows the standard electrode potentials $E^{\ominus}$ for the interconversion of the M^{3+} and M^{2+} ions. The negative $E^{\ominus}$ values for titanium, vanadium, and chromium show that M^{2+} is a relatively strong reductant in each case. We need to use an even stronger reductant such as zinc metal ($E^{\ominus} = -0.76$ V) to reduce M^{3+} to M^{2+} for titanium, vanadium, and chromium.

We can explain the large increase in the $E^{\ominus}$ value for manganese and then decrease for iron, seen in Figure 14.1. Both Mn^{2+} and Fe^{3+} have five electrons in the d subshell; this *half-filled* subshell is particularly stable. For manganese the more common ion is therefore Mn^{2+}. For iron, however, both Fe^{2+} and Fe^{3+} are easily obtained. The more stable ion for cobalt is Co^{2+}, as shown in Figure 14.1 by its very high $E^{\ominus}$ value.

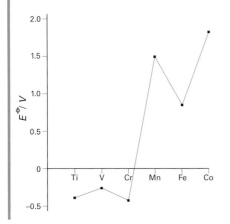

FIGURE 14.1 The standard electrode potentials $E^{\ominus}$ for reduction of M^{3+} to M^{2+} for Ti to Co. (In aqueous solution, the M^{3+} ion is very unstable for both Ni and Cu.)

- **Transition metals in high oxidation states become more oxidizing.**
  The two most important examples are manganate(VII) ion and dichromate(VI) ion. We can use either in acidic solution to oxidize both

inorganic ions such as $Fe^{2+}(aq)$ and organic molecules such as ethanol (Section 21.4).

- The **intensely purple** aqueous solution containing the **manganate(VII) ion MnO_4^-** is reduced to an almost **colourless** solution containing the ion Mn^{2+}:

$$MnO_4^-(aq) + 8H^+(aq) + 5e^- \rightarrow Mn^{2+}(aq) + 4H_2O(l)$$

- We do not need to use an indicator in **redox titrations** as the appearance of the purple colour indicates when the manganate(VII) ion is in excess.
- The best acid to use for manganate(VII) titrations is dilute sulfuric acid as manganate(VII) ion can oxidize the chloride ion in dilute hydrochloric acid.

- The **orange** aqueous solution containing the **dichromate(VI) ion $Cr_2O_7^{2-}$** is reduced to a **green** solution containing the ion Cr^{3+}:

$$Cr_2O_7^{2-}(aq) + 14H^+(aq) + 6e^- \rightarrow 2Cr^{3+}(aq) + 7H_2O(l)$$

- Being a weaker oxidizing agent than manganate(VII) ion, dichromate(VI) ion can be used with dilute hydrochloric acid.
- Taking the Next Step 14.2 explains how to reduce a transition metal ion.
- Taking the Next Step 14.3 explains how to oxidize a transition metal ion.

> The dichromate(VI) ion is also called the **dichromate ion**. At university, the manganate(VII) ion is frequently called the **permanganate ion** (A Deeper Look 13.1).

> The $E^{\ominus}$ value for $Cr_2O_7^{2-}/Cr^{3+}$ (+1.33 V) is just *smaller* than the $E^{\ominus}$ value for Cl_2/Cl^- (+1.36 V) whereas the $E^{\ominus}$ value for MnO_4^-/Mn^{2+} (+1.51 V) is significantly larger.

14

TAKING THE NEXT STEP 14.2
How can we reduce a transition metal ion?

We can **reduce a transition metal ion** by making the solution *acidic* and then adding a *reductant* such as zinc metal. An effective demonstration is to reduce ammonium vanadate(V) using zinc and HCl. Vanadium passes through the oxidation states IV and III quite quickly before eventually forming oxidation state II. The ions present in these oxidation states and their colours are:

$$Ox(V) = +5 \quad VO_2^+(aq) \quad yellow$$
$$Ox(V) = +4 \quad VO^{2+}(aq) \quad blue$$
$$Ox(V) = +3 \quad V^{3+}(aq) \quad green$$
$$Ox(V) = +2 \quad V^{2+}(aq) \quad violet.$$

Figure 14.2 illustrates the violet solution containing V^{2+} ions.

FIGURE 14.2 Zinc amalgam has reduced acidified vanadate(V) ions to vanadium(II) ions, V^{2+}, which are violet coloured. The reaction takes about a week.

TAKING THE NEXT STEP 14.3
How can we oxidize a transition metal ion?

The stability of the higher oxidation states for all the metals can be significantly increased in alkaline solution. (See A Deeper Look 14.1 for more details.) We can therefore **oxidize a transition metal ion** by making the solution *alkaline* and then adding an *oxidant* such as hydrogen peroxide; oxygen from the air may also be effective.

Aqueous chromium(III) ions in excess sodium hydroxide form a green solution containing the ion $[Cr(OH)_6]^{3-}$ (Section 15.7); we can oxidize using hydrogen peroxide to form a yellow solution containing the chromate(VI) ion $CrO_4^{2-}(aq)$.

The iron(II) hydroxide precipitate formed when we add sodium hydroxide to aqueous iron(II) ions is green initially, but in this alkaline solution exposure to oxygen at the surface turns the colour red-brown. Figure 14.3 illustrates this oxidation.

FIGURE 14.3 Green iron(II) hydroxide forms when aqueous sodium hydroxide is added to aqueous iron(II) ions. The precipitate oxidizes on contact with air to red-brown iron(III) hydroxide.

A DEEPER LOOK 14.1

What effect has pH on electrode potentials?

A Deeper Look 8.2 explained the fact that electrode potentials vary with pH. As the pH increases, higher oxidation states become more stable. Figure 14.4 shows this behaviour for iron; this is an example of a **Pourbaix diagram**. Note that at pH 0 both Fe^{2+} ions and Fe^{3+} ions have a considerable range of electrode potentials where they are stable. At pH 14, iron(II) hydroxide has a very narrow range of stability compared with iron(III) hydroxide.

This Pourbaix diagram also allows us to work out the solubility product K_{sp} (A Deeper Look 12.1, treated as a thermodynamic equilibrium constant, Section 6.1) for iron(II) hydroxide very quickly:

$$K_{sp} = [Fe^{2+}][OH^-]^2$$

The vertical line linking the $Fe^{2+}(aq)$ and $Fe(OH)_2(s)$ areas occurs at a pH of 6.5.

So $[H^+] = 10^{-6.5}$ and therefore $[OH^-] = 10^{-14}/10^{-6.5} = 10^{-7.5}$ and $K_{sp} = 1 \times (10^{-7.5})^2 = 1 \times 10^{-15}$.

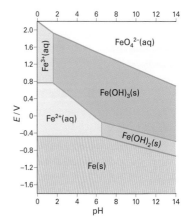

FIGURE 14.4 A **Pourbaix diagram** shows how the electrode potential varies with pH (the above figure applies to 1 mol dm^{-3} solutions). The values at pH 0 correspond to the standard electrode potentials. The Nernst equation (A Deeper Look 8.2) describes the slopes of the diagonal lines.

14.3 CATALYTIC BEHAVIOUR

- **Transition metals are often used as catalysts**; the metals themselves, their ions, and their compounds are all used. Examples are as follows:
  - Titanium(III) chloride is the catalyst in the Ziegler–Natta polymerization of ethene (Section 25.2).
  - Vanadium(V) oxide is the catalyst in the Contact process for making sulfuric acid. (We look in detail at this reaction in Section 14.5.)
  - Manganese(IV) oxide catalyses the decomposition of hydrogen peroxide.

FIGURE 14.5 Cobalt(II) ions catalyse the reaction between hydrogen peroxide and Rochelle salt. The colour changes to green, the colour of cobalt(III) ions, before reverting to pink, the colour of cobalt(II) ions, at the end of the reaction.

- Metallic iron is the catalyst in the Haber–Bosch process for making ammonia (Section 6.3).
- Figure 14.5 illustrates that cobalt(II) ions catalyse the reaction between hydrogen peroxide and Rochelle salt.
- Metallic nickel (or palladium or platinum) is the catalyst in the hydrogenation (Section 14.5) of vegetable oils to form margarine.
- Catalytic converters (Section 14.5) contain platinum and rhodium metals.

• We can classify catalysts as either

  – **homogeneous catalysts**, those in the same phase as the reactants, or
  – **heterogeneous catalysts**, those in a different phase from the reactants.

14.4 HOMOGENEOUS CATALYSTS

• **Homogeneous catalysts almost always depend on the variable oxidation states available to a transition metal.**

  – For example, the reaction between iodide ion and the very strongly oxidizing persulfate ion ($S_2O_8{}^{2-}$) is slow because of the mutual repulsion between the two negatively charged ions:
    $$S_2O_8{}^{2-}(aq) + 2I^-(aq) \rightarrow 2SO_4{}^{2-}(aq) + I_2(aq)$$
    ○ Iron(II) ions can catalyse this reaction by being oxidized by the persulfate to iron(III) ions, which are themselves then reduced back to iron(II) ions by the iodide ions.
    ○ $S_2O_8{}^{2-}(aq) + 2Fe^{2+}(aq) \rightarrow 2SO_4{}^{2-}(aq) + 2Fe^{3+}(aq)$
    ○ $2Fe^{3+}(aq) + 2I^-(aq) \rightarrow 2Fe^{2+}(aq) + I_2(aq)$

The $E^\ominus$ value for Fe^{3+}/Fe^{2+} (+0.76 V) lies between the $E^\ominus$ values for $S_2O_8{}^{2-}/SO_4{}^{2-}$ (+2.01 V) and I_2/I^- (+0.54 V).

- Homogeneous catalysis occurs between manganese(II) ions and manganate(VII) ions. This means that the redox titration between manganate(VII) ions and ethanedioate (oxalate) ions $C_2O_4^{2-}$ becomes faster as the reaction proceeds. We call this **autocatalysis**, as a product of the reaction acts as a catalyst for the reaction:
  - $4Mn^{2+}(aq) + MnO_4^-(aq) + 8H^+(aq) \rightarrow 5Mn^{3+}(aq) + 4H_2O(l)$
  - $2Mn^{3+}(aq) + C_2O_4^{2-}(aq) \rightarrow 2Mn^{2+}(aq) + 2CO_2(g)$

14.5 HETEROGENEOUS CATALYSTS

Adsorption occurs **on** a surface: **ab**sorption occurs if the reactants penetrate **into** the bulk of the catalyst.

- Heterogeneous catalysts work as follows. The reactants diffuse to the surface, **adsorb** onto the surface, **react** on the surface, **desorb** off the surface, and diffuse away. The most common examples involve solid metals catalysing the reaction of gases.
- Figure 14.6 shows how ethene and hydrogen react to form ethane on a nickel surface.
- Figure 14.7 shows how the Fischer–Tropsch process works to manufacture liquid hydrocarbons from carbon monoxide and hydrogen.

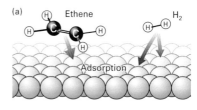

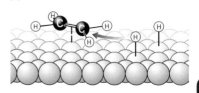

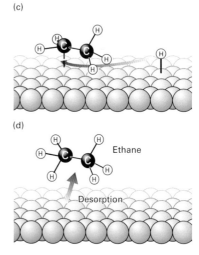

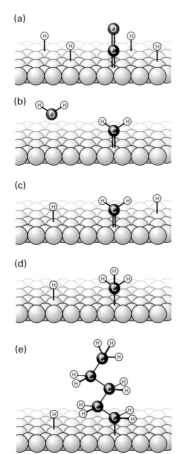

FIGURE 14.7 The mechanism that occurs over cobalt catalysts in the Fischer–Tropsch (FT) SASOL process:
(a) CO and H atoms on the Co surface.
(b) $=CH_2$ on the surface and H_2O in the gas phase.
(c) An H atom on the surface reacts with $=CH_2$ on the surface to produce
(d) $-CH_3$, which then reacts further.
(e) The product (pentane here) forms on the surface by reaction with the H atom and then desorbs.

FIGURE 14.6 (a) Hydrogen and ethene adsorb onto the metal surface by donating electron density into vacant d orbitals on the catalyst metal atoms. (b) and (c) Hydrogen atoms (radicals) interact with the electron density in the ethene double bond, and bond with the carbon atoms. (d) The product ethane desorbs from the surface of the catalyst.

14

FIGURE 14.8 The reaction mechanism for the catalytic reduction of nitrogen by hydrogen on an iron surface. This diagram shows the three stages: (a) inward diffusion of reactants and (b) attachment to the catalyst surface; followed by reaction to form products; and (c) desorption and outward diffusion of products.

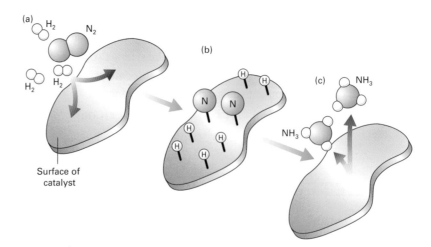

The presence of precious metals in catalytic converters makes them a target for thieves.

- The iron catalyst in the Haber–Bosch process (Section 6.3) speeds up the reaction because the nitrogen molecule with its very strong triple bond adsorbs onto the surface as individual atoms. This process breaks the strongest bond that needs to be broken, which makes subsequent reaction much easier. Figure 14.8 shows this **dissociative adsorption**.

- **The variable oxidation states available to a transition metal are often critical for the catalytic process.** A good example in the case of a heterogeneous catalyst is the use of vanadium(V) oxide V_2O_5 in the Contact process (Section 14.3). The vanadium interchanges between oxidation states V and IV. The reaction to be catalysed,

  $2SO_2 + O_2 \rightarrow 2SO_3$, occurs as follows:
  - $2SO_2 + 2V_2O_5 \rightarrow 2SO_3 + 4VO_2$
    - Ox(S) changes from +4 to +6; Ox(V) changes from +5 to +4
  - $4VO_2 + O_2 \rightarrow 2V_2O_5$
    - Ox(V) changes from +4 to +5, Ox(O) changes from 0 to −2

- **Catalytic converters** in car exhausts use platinum metal with about 10% rhodium. The metals are spread on a **catalyst support** (a ceramic honeycomb), which provides physical support and increases the surface area.

- **Poisoning** can reduce the effectiveness of heterogeneous catalysts, as they can become coated, reducing the surface area available for adsorption. One example is the poisoning of a catalytic converter with lead; hence the need for unleaded petrol.

- An **inhibitor** is a chemical that reduces the rate or effectiveness of a chemical or biochemical process. The activity of many enzymes can be inhibited by the binding of specific small molecules and ions. See also Section 16.3.

TRANSITION METALS 2

15.1 COMPLEX IONS

- Figure 15.1 shows that a **complex ion** is a central metal ion surrounded by ligands (see next bullet point).
  - **Coordinate bonding** (Section 2.5) holds the complex ion together.
  - The **coordination number** of a complex ion is the number of coordinate bonds made to the central metal ion.
- A **ligand** can donate at least one electron pair to the central metal ion.
  - Ligands are Lewis bases (Section 7.6); the central metal ion is a Lewis acid.
  - We can classify ligands as unidentate, bidentate, or multidentate.
- **Unidentate (monodentate) ligands** donate one electron pair and include the uncharged molecules water and ammonia plus hydroxide, chloride, and cyanide ions.
- Note how this list parallels very closely the list of the most common types of nucleophiles (Section 20.2). Nucleophiles are also electron-pair donors.
- The two most common shapes of complex ions with unidentate ligands are **octahedral** (when six ligands are attached) and **tetrahedral** (when four ligands are attached).

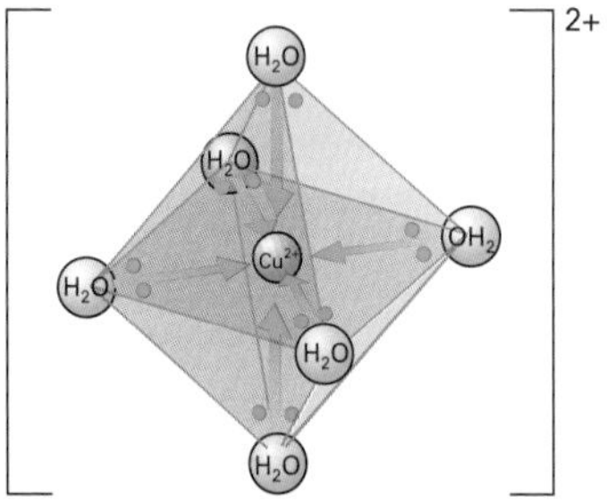

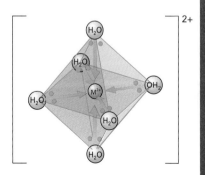

FIGURE 15.1 The complex ion $[Cu(H_2O)_6]^{2+}$. The oxygen atom in each water molecule acts as a Lewis base (Section 7.6), donating an electron pair to the central metal ion. Each such electron pair constitutes a coordinate bond, shown by the blue arrow.

15.2 OCTAHEDRAL COMPLEXES

- **Aqua complexes** typically have six water molecules as ligands: all the first-row transition metals form an octahedral **hexaaqua ion** of formula $[M(H_2O)_6]^{2+}$, as shown in Figure 15.2.
  - Because it is the oxygen that donates a lone pair to form the coordinate bond, you will often see the formula of the ion written as $[M(OH_2)_6]^{2+}$.
  - The common hexaaqua ions in oxidation state II are coloured (Section 15.9).

FIGURE 15.2 The octahedral hexaaqua ion $[M(H_2O)_6]^{2+}$.

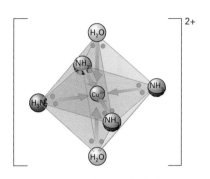

FIGURE 15.3 The octahedral complex ion $[Cu(NH_3)_4(H_2O)_2]^{2+}$.

However, Hermann Jahn and Edward Teller explained in 1937 that a complex with 9 d electrons on the central ion, as is the case for Cu^{2+}, would not have a perfectly symmetrical shape but would necessarily be distorted.

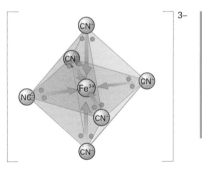

FIGURE 15.4 The octahedral complex ion $[Fe(CN)_6]^{3-}$: the carbon of the cyanide ion donates an electron pair.

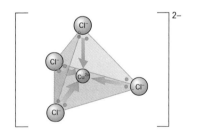

FIGURE 15.5 The tetrahedral complex ion $[CuCl_4]^{2-}$: each chloride ion donates an electron pair.

We can use **cobalt(II) chloride paper** as a common colour test for water, as the presence of water changes the paper from blue to pink.

- Some of the first row transition metals also form an octahedral hexaaqua ion of formula $[M(H_2O)_6]^{3+}$, e.g. $[Fe(H_2O)_6]^{3+}$. (See Figures 15.14 and 15.18.)

- **Ammine complexes** also typically have six ammonia molecules as ligands: some of the first row transition metals form an octahedral **hexaammine ion** of formula $[M(NH_3)_6]^{2+}$. Formation from the hexaaqua ion is an example of a **ligand substitution** (ligand exchange) reaction, in which one ligand replaces another.

  - Ammine complexes are more common towards the end of the first row of the transition metals, especially $[Co(NH_3)_6]^{2+}$ and $[Ni(NH_3)_6]^{2+}$, which form when we add excess aqueous ammonia to a solution containing the metal(II) ions (Section 15.8).

  - Copper forms the most famous of all ammine complexes, which produces a deep blue-violet solution. Figure 15.3 shows that its formula is rather unexpected, being $[Cu(NH_3)_4(H_2O)_2]^{2+}$, which is octahedral. Figure 15.13 illustrates both this ammine complex and the original hexaaqua ion.

Taking the Next Step 15.1 discusses **cyano complexes**.

TAKING THE NEXT STEP 15.1

Cyano complexes

Cyano complexes also typically have six cyanide ions as ligands. The most important examples are the hexacyanoferrate(II) ion $[Fe(CN)_6]^{4-}$ and the hexacyanoferrate(III) ion $[Fe(CN)_6]^{3-}$, which is shown in Figure 15.4. See also Figure 15.22.

15.3 TETRAHEDRAL COMPLEXES

- **Chloro complexes of the first-row transition metals are typically tetrahedral**, as shown in Figure 15.5, because the larger, charged chloride ions repel each other. Taking the Next Step 15.2 explains how to write the formulae for chloro complexes.

- Adding concentrated hydrochloric acid to a solution containing the hexaaqua ion usually forms the **tetrachloro ion**, as a result of ligand substitution. Figure 15.5 shows the tetrachlorocuprate(II) ion.

  - $[Cu(H_2O)_6]^{2+}(aq) + 4Cl^-(aq) \rightarrow [CuCl_4]^{2-}(aq) + 6H_2O(l)$

    ○ the colour of the solution changes from blue to yellow-green (see Figure 15.13).

  - $[Co(H_2O)_6]^{2+}(aq) + 4Cl^-(aq) \rightarrow [CoCl_4]^{2-}(aq) + 6H_2O(l)$

    ○ the colour of the solution changes from pink to blue.

We can reverse these reactions by adding excess water to the tetrachloro ions.

How do we write the formulae for chloro complexes and oxoanions?

The formula of a chloro complex traditionally uses square brackets round the complex ion, in exactly the same way that we use square brackets for aqua and ammine complexes. However, aqua and ammine complexes involve species that contain more than one element in the ligand and as such round brackets *must* be used round the molecule itself plus square brackets round the complex ion.

The logically identical situation when the ligands are oxide ions (which like chloride ions involve only one element) traditionally uses *no* brackets round the ion, which we call an **oxoanion**. So the usual formula for the hexachloro complex of chromium(III) is $[CrCl_6]^{3-}$, yet for the chromate(VI) ion the usual formula is CrO_4^{2-}.

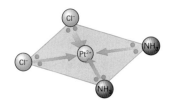

FIGURE 15.6 The square planar complex ion $[Ni(CN)_4]^{2-}$.

15.4 MORE UNUSUAL SHAPES FOR COMPLEXES

- Four ligands can also occasionally adopt a **square planar** shape. The most important group to show this behaviour is Group 10 (nickel, palladium, and platinum) when they are in oxidation state II: these ions have 8 d electrons. Figure 15.6 shows one example: the tetracyanonickelate(II) ion $[Ni(CN)_4]^{2-}$.

  - Figure 15.7 shows a very important example of a square planar complex: the widely prescribed anti-cancer drug **cisplatin** $[Pt(NH_3)_2Cl_2]$. Note that this is the *cis* isomer and that a *trans* isomer (Section 16.3) also exists.

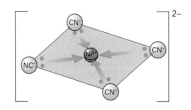

FIGURE 15.7 The *cis* isomer of the square planar complex $[Pt(NH_3)_2Cl_2]$.

- When there are only two ligands, the shape is **linear**. Figure 15.8 shows the diamminesilver(I) ion $[Ag(NH_3)_2]^+$, which is found in Tollens' reagent (Section 22.2).

FIGURE 15.8 The linear complex ion $[Ag(NH_3)_2]^+$.

15.5 BIDENTATE LIGANDS

- **Bidentate ligands** contain two atoms, both of which can donate an electron pair. One important example is 1,2-diaminoethane, which is also called ethylenediamine (shortened to **en**): $NH_2CH_2CH_2NH_2$ in which the two nitrogen atoms each donate their lone pair. Complexes can form with one, two, or three en ligands.

- When two en ligands are present, **isomers** can exist, called *cis* and *trans*. The *cis* isomer is particularly interesting as it can exist in two chiral forms (Section 16.3). Figure 15.9 shows both isomers.

Another bidentate ligand is **ethanedioate (oxalate) ion** $(COO^-)_2$.

15

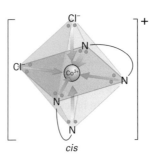
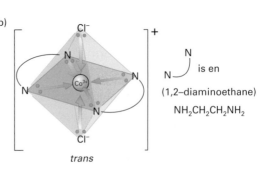

(a) *cis*   *cis*   (b) *trans*

N—
N— is en
(1,2–diaminoethane)
$NH_2CH_2CH_2NH_2$

FIGURE 15.9 (a) The *cis* isomer of $[Co(en)_2Cl_2]^+$ and its mirror image are not superimposable. They form a pair of optical isomers (Figure 16.8). (b) The *trans* isomer has a plane of symmetry (Figure 16.13). It is superimposable on its mirror image.

15.6 MULTIDENTATE LIGANDS

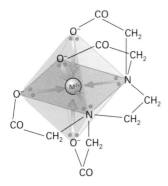

FIGURE 15.10 The hexadentate ligand $EDTA^{4-}$ forming a complex ion with a metal(II) ion.

• **Multidentate (polydentate) ligands** contain many atoms capable of donating an electron pair. An important example is the hexadentate ligand **$EDTA^{4-}$**, in which four oxygens and two nitrogens donate lone pairs.
  – Figure 15.10 shows the structure of the complex.
  – When you are drawing the structure, it is easiest to draw in the atoms forming the coordinate bonds before trying to link these atoms together.

• Figure 15.11 shows the particularly important tetradentate ligand **haem**, present in the haemoglobin molecule.
  – Haemoglobin transfers oxygen in the blood: because carbon monoxide bonds more strongly than oxygen does to the iron ion at the centre of haemoglobin, CO can form carboxyhaemoglobin and hence cause asphyxiation.

FIGURE 15.11 The electrostatic potential map of the haem ring.

The name 'chelate' comes from the Greek for 'crab's claw'.

The formula of the metal(III) hydroxide precipitate is uncertain. It is traditionally written as $M(OH)_3$, but $MO(OH)$ may be more accurate.

• Bidentate and especially multidentate ligands form very stable complexes, which we call the **chelate effect**. The chelate effect is caused by the large entropy increase (Section 9.1) that occurs as more particles are released into the solution as the ligand substitution occurs.

> We use chelating ligands to treat heavy metal poisoning, by mercury or lead, for example.

15.7 REACTION WITH AQUEOUS SODIUM HYDROXIDE

• Taking the Next Step 15.3 explains why metal(III) ions are acidic (and therefore react with alkalis).

• The usual first result of adding dilute aqueous sodium hydroxide to a solution of a metal ion is the **precipitation of the metal hydroxide**. Figure 15.12 shows examples of various insoluble metal hydroxides. A typical equation for the reaction is as follows:
  – $[Fe(H_2O)_6]^{3+}(aq) + 3OH^-(aq) \rightarrow Fe(OH)_3(s) + 6H_2O(l)$

FIGURE 15.12 Precipitates of various metal hydroxides.

- We can use the **colours** of some of the hydroxides to identify the metal ions:
  - $Cr(OH)_3$ green precipitate
  - $Fe(OH)_3$ red-brown precipitate
  - $Fe(OH)_2$ green precipitate (turns red-brown on standing in air, Figure 14.3)
  - $Cu(OH)_2$ blue precipitate

TAKING THE NEXT STEP 15.3

Why are metal(III) aqua ions acidic?

Metal(III) aqua ions have a high charge density (Section 2.7). This high charge density attracts electron density from the oxygen atom of a water ligand (Figure 15.1). This weakens the O–H bond, making it possible for a water ligand to lose a proton. Proton loss produces a complex ion with five water ligands and one hydroxide ligand:

$$[M(H_2O)_6]^{3+} + H_2O \rightleftharpoons H_3O^+ + [M(H_2O)_5OH]^{2+}$$

We can then write equilibria showing loss of two and then three protons.
Addition of sodium hydroxide to this mixture of species will shift all the equilibria to the right, causing the metal(III) hydroxide to precipitate.

The acidity of aqueous iron(III) ions is similar to that of ethanoic acid.

One other consequence of this acidity equilibrium is that the $[Fe(H_2O)_5OH]^{2+}$ ion is responsible for the yellow-green colour (see Figure 15.14) of aqueous iron(III) ions. The hydroxo complex is much more intensely coloured, even though the hexaaqua ion still has a much higher concentration in the solution.

Metal(II) ions are less acidic because the charge density of a metal(II) ion is significantly lower than that of a metal(III) ion. However, once again, addition of sodium hydroxide will shift all the equilibria to the right, causing the metal(II) hydroxide to precipitate.

Amphoteric behaviour

- A small number of hydroxides react further when we add **excess NaOH(aq)**. The green solid chromium(III) hydroxide reacts ('**redissolves**') to form a green solution containing the hexahydroxochromate(III) ion:

$$Cr(OH)_3(s) + 3OH^-(aq) \rightarrow [Cr(OH)_6]^{3-}(aq)$$

We describe any hydroxide that behaves in this way as **amphoteric**.

A Deeper Look 15.1 describes another pair of metals (which are not transition metals) that form amphoteric hydroxides.

A DEEPER LOOK 15.1

Which other hydroxides are amphoteric?

The white solid aluminium hydroxide 'redissolves' to form a colourless solution containing the tetrahydroxoaluminate ion:

$$Al(OH)_3(s) + OH^-(aq) \rightarrow [Al(OH)_4]^-(aq)$$

The white solid zinc hydroxide 'redissolves' to form a colourless solution containing the tetrahydroxozincate ion:

$$Zn(OH)_2(s) + 2OH^-(aq) \rightarrow [Zn(OH)_4]^{2-}(aq)$$

We can distinguish $Zn(OH)_2$ from $Al(OH)_3$ because the former 'redissolves' in excess aqueous ammonia, in a similar way that $Co(OH)_2$, $Ni(OH)_2$, and $Cu(OH)_2$ do (which we describe in Section 15.8). However, the complex ion formed is $[Zn(NH_3)_4]^{2+}(aq)$.

15.8 REACTION WITH DILUTE AQUEOUS AMMONIA

- **Dilute aqueous ammonia is an alkali** (Section 7.1) and so it produces hydroxide ions. Therefore the first result on adding dilute aqueous ammonia, as for dilute sodium hydroxide, is the **precipitation of the metal hydroxide**.

- A small number of transition metal hydroxides react further when we add *excess* **aqueous ammonia**.
  - The blue-green precipitate cobalt(II) hydroxide reacts ('**redissolves**') to form a straw-coloured solution containing a **hexaammine** complex ion:

    $$Co(OH)_2(s) + 6NH_3(aq) \rightarrow [Co(NH_3)_6]^{2+}(aq) + 2OH^-(aq)$$

  - The green precipitate nickel(II) hydroxide reacts to form a violet-coloured solution containing a hexaammine complex ion:

    $$Ni(OH)_2(s) + 6NH_3(aq) \rightarrow [Ni(NH_3)_6]^{2+}(aq) + 2OH^-(aq)$$

  - The blue precipitate copper(II) hydroxide reacts to form a dark blue-violet solution containing the ion $[Cu(NH_3)_4(H_2O)_2]^{2+}$ (Figure 15.3). Figure 15.13 illustrates this reaction.

    $$Cu(OH)_2(s) + 4NH_3(aq) + 2H_2O(l) \rightarrow [Cu(NH_3)_4(H_2O)_2]^{2+}(aq) + 2OH^-(aq)$$

15

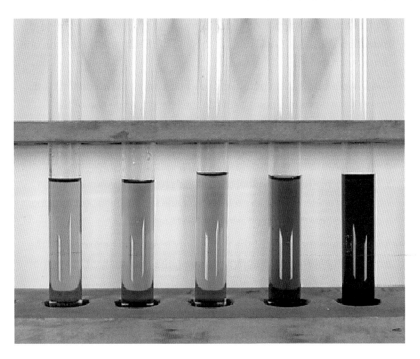

FIGURE 15.13 Copper(II) complexes change colour depending on the ligand. The tubes on the left have concentrated and dilute hydrochloric acid added. The tubes on the right have dilute and concentrated aqueous ammonia added.

15.9 FORMATION OF COLOURED IONS

- The colours of the common hexaaqua ions in oxidation state II are
    - $[Mn(H_2O)_6]^{2+}$ pale pink
    - $[Fe(H_2O)_6]^{2+}$ green
    - $[Co(H_2O)_6]^{2+}$ pink
    - $[Ni(H_2O)_6]^{2+}$ green
    - $[Cu(H_2O)_6]^{2+}$ blue

Figure 15.14 illustrates four of these ions along with other coloured ions.

Think how often you have seen science labs in films or TV shows with a container full of a blue solution!

15

FIGURE 15.14 Common transition metal ions in aqueous solution. Left to right: VO_4^{3-}, V^{3+}, Cr^{2+}, Cr^{3+}, $Cr_2O_7^{2-}$, Mn^{2+}, MnO_4^-, Fe^{3+}, Co^{2+}, Ni^{2+}, Cu^{2+}.

(a)   (b)

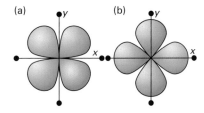

FIGURE 15.15 In an octahedral complex, (a) three d orbitals point *between* the ligands (represented by the black dots), which lie along the axes, and (b) two d orbitals point *directly at* the ligands.

- We call the most important mechanism for producing colour a **d-to-d transition**. A Deeper Look 15.4 (at the end of this chapter) explains one other mechanism.

- The five d orbitals in an octahedral complex come in two distinct sets: three of the d orbitals point *between* the ligands, while two of the d orbitals point *directly at* the ligands, as shown in Figure 15.15.

- The energy of these two sets in the presence of the six ligands in an octahedral complex ion is different, causing what we call a **ligand field splitting**, as shown in Figure 15.16. (Remember that the d orbitals have the same energy in an isolated ion or atom.) We can explain this ligand field splitting best by using molecular orbital theory; see Taking the Next Step 15.4.

- An electron in the lower of the two energy levels can be excited to the higher energy level, in so doing absorbing the frequency of light f that corresponds exactly to the energy gap ΔE:

$$\Delta E = hf$$

Figure 15.17 shows this absorption.

- This frequency is removed from white light, and we see the complementary colour. Figure 15.18 illustrates that for $[Ti(H_2O)_6]^{3+}$ absorption is in the green, so red and blue light is transmitted and the sample appears purple.

- The colour of complex ions depends on the nature and oxidation state of the central metal ion as well as on the number and nature of the ligands. See, for example, Figures 15.13 and 15.14.

15

Octahedral complex: two d orbitals at higher energy

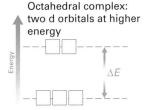

FIGURE 15.16 The d orbitals in an octahedral complex split into two energy levels.

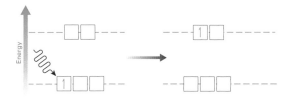

FIGURE 15.17 Absorption of a photon excites the electron in the lower energy level into the higher energy level. The corresponding colour of light is removed from the incident white light.

(a)

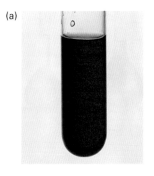

(b)

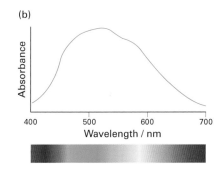

FIGURE 15.18 (a) A solution containing the complex ion $[Ti(H_2O)_6]^{3+}(aq)$. (b) The absorption spectrum of $[Ti(H_2O)_6]^{3+}(aq)$.

TAKING THE NEXT STEP 15.4

How does the ligand field split the d orbitals?

The best way to explain the effect of an octahedral ligand field on the d orbitals is to use molecular orbital theory. Two of the d orbitals—those lying along the axes—have the correct symmetry to interact with the ligands. As usual, bonding and antibonding orbitals form. Figure 15.19 shows one of the antibonding orbitals.

This splits the five d orbitals into two groups: a set of three (which will be labelled at university as t_{2g}) that have no interaction with the ligands, and a set of two (which will be labelled at university as e_g) that do interact. These two orbitals are the antibonding orbitals which lie *higher* in energy than the t_{2g} d orbitals, as shown in Figure 15.16.

A Deeper Look 15.2 explains how d-orbital splitting gives rise to low-spin and high-spin complexes.

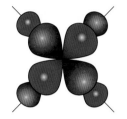

FIGURE 15.19 One of the two antibonding MOs formed from the d orbitals overlapping with the ligands in an octahedral complex ion.

A DEEPER LOOK 15.2

What are high-spin and low-spin complexes?

Given that splitting in an octahedral ligand field produces a set of three energy levels lying below two, there are two possibilities for the electronic structure of transition metal ions with between four and seven d electrons.

One possibility is to have as many unpaired electrons as possible, which produces a **high-spin complex**. If the gap between the two energy levels is larger, pairing of the electrons in the lower energy level produces a **low-spin complex**. Figure 15.20 shows these two possibilities. The formation of a low-spin complex is of particular importance in the role the iron(II) ion plays in breathing, see A Deeper Look 15.3.

Look again at Figure 15.20. It shows that for all the examples except for low-spin d^6 (as is the case for low-spin Fe^{2+}, A Deeper Look 15.3) there is at least one unpaired electron. Taking the Next Step 2.1 explains that any unpaired electron causes **paramagnetism**.

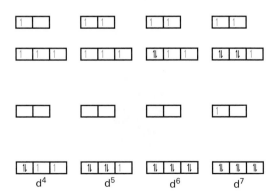

FIGURE 15.20 For ions with between four and seven d electrons, there can be either high-spin (top two lines) or low-spin complexes (bottom two lines).

(a)

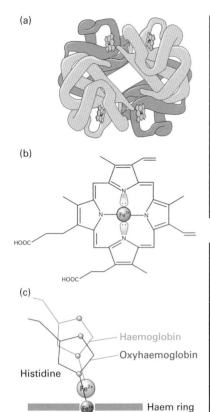

(b)

(c)

Histidine

Haemoglobin
Oxyhaemoglobin

Haem ring

FIGURE 15.21 (a) Haemoglobin. (b) The Fe^{2+} ion forms bonds with four nitrogen atoms in the haem ring and a fifth bond with histidine, present in another section of the protein. (c) When oxygen binds to haemoglobin, the high-spin Fe^{2+} ion becomes a low-spin Fe^{2+} ion, which is smaller. This smaller ion can fit into the space at the centre of the haem ring. This has a significant knock-on effect on the rest of the haemoglobin molecule.

A DEEPER LOOK 15.3

What happens to haemoglobin's iron ion during breathing?

The iron ion in haemoglobin is Fe^{2+}, which has $8 - 2 = 6$ electrons in the d orbitals. Approximating the ligand field around the ion as octahedral and assuming that the energy gap is small, this creates a high-spin d^6 ion. Adding an extra ligand (when the nearer oxygen atom of the O_2 molecule binds to haemoglobin to form oxyhaemoglobin) increases the ligand field, and the iron ion becomes a *low-spin* d^6 ion. The low-spin ion, being smaller, can then fit into the space at the centre of the haem ring, as shown in Figure 15.21. This has a significant knock-on effect on the rest of the haemoglobin molecule.

A DEEPER LOOK 15.4

Coloured ions by a charge-transfer transition

One other important mechanism for creating colour is a **charge-transfer transition**. This occurs when an electron transfers from the *ligand* to the central metal ion. The most important examples of this mechanism are the deep purple of manganate(VII) ions and the orange of dichromate(VI) ions (Figure 15.14).

Figure 15.22 illustrates the highly insoluble pigment **Prussian blue**. We can form this by adding iron(II) ions to potassium hexacyanoferrate(III) or iron(III) ions to potassium hexacyanoferrate(II): note that both oxidation states of iron must be present. The other essential ingredient is cyanide ions to act as ligands. Cyanide's presence in this pigment caused Joseph Gay-Lussac to give cyanide its name (from the Greek word for 'blue').

Prussian blue was used for traditional blueprints, as well as for Prussian army uniforms from the eighteenth to the early twentieth centuries. The formula of Prussian blue is $Fe^{III}_4[Fe^{II}(CN)_6]_3$.

FIGURE 15.22 The formation of Prussian blue. Here, aqueous hexacyanoferrate(III) ions $[Fe(CN)_6]^{3-}$ are being added to aqueous iron(II) sulfate.

INTRODUCTION TO ORGANIC CHEMISTRY

Organic chemistry involves the study of the compounds formed by **carbon**. Living organisms are mainly composed of carbon compounds.

16.1 FORMULAE FOR ORGANIC MOLECULES

Section 3.1 gives the definitions of empirical formula and molecular formula.

- The **structural formula** shows exactly which atoms are bonded together.
  - A **condensed structural formula** expresses this in an unambiguous but rapid form while a **displayed formula** draws out all the bonds explicitly, as shown in Figure 16.1. (It is easy to forget to include the O−H bond in an alcohol.)

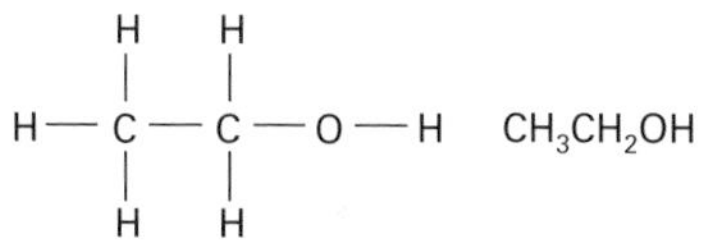

FIGURE 16.1 The displayed formula of ethanol and its condensed structural formula.

- Figure 16.2 shows some **skeletal formulae**, which strip away much of the detail to focus only on the carbon skeleton plus any functional groups present.

- Figure 16.3 shows how to draw a molecule in three dimensions.
  - Lines drawn like normal bonds lie in the plane of the paper.
  - **Bold wedged lines** represent bonds pointing forwards out of the plane of the paper.
  - **Dashed (cross-hatched) lines** represent bonds pointing backwards behind the plane of the paper.

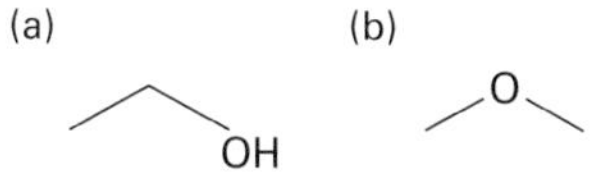

FIGURE 16.2 Skeletal formulae: (a) ethanol and (b) methoxymethane.

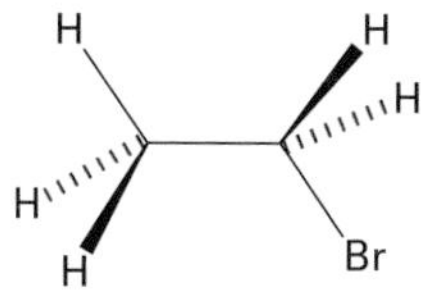

- A **functional group** is the group of atoms that gives the molecule its characteristic chemical properties. All alcohols, for example, must have at least one hydroxyl (−OH) group.

- A **homologous series** is a collection of molecules with the same functional group differing only in the number of carbon atoms present: each successive member of the series has one more CH_2 group. Homologous series have a **general formula**, **similar chemical properties**, and **gradually changing physical properties**. The simplest homologous series is the **alkanes** (C_nH_{2n+2}). Other series include the **alkenes** (C_nH_{2n}) and the **alcohols** ($C_nH_{2n+2}O$).

FIGURE 16.3 Drawing a molecule in 3D.

16.2 IUPAC NOMENCLATURE

- We name the first six alkanes methane (CH_4), ethane (C_2H_6), propane (C_3H_8), butane (C_4H_{10}), pentane (C_5H_{12}), and hexane (C_6H_{14}). We can make an **alkyl group** (general symbol **R**) by removing one hydrogen atom from an alkane; see Taking the Next Step 16.1.

> ### TAKING THE NEXT STEP 16.1
> #### What abbreviations are used for alkyl groups?
>
> Abbreviations used at university for the most common alkyl groups are as follows:
>
> **Me** stands for methyl $-CH_3$
>
> **Et** stands for ethyl $-CH_2CH_3$
>
> **Pr** stands for propyl $-CH_2CH_2CH_3$
>
> **iPr** stands for *iso*propyl $-CH(CH_3)_2$
>
> **tBu** stands for *tertiary*butyl $-C(CH_3)_3$

- We base the **IUPAC names** for other molecules on those for the alkanes:
  - Find the longest continuous carbon chain (the **spine**) containing the functional group (if there is one).
  - Name the functional group present.
  - Name any side chains.
  - Number the position of the functional group (and the side chains): choose the *lower number at the first point of difference* if there is a choice.
  - If more than one identical group is present, use the multipliers di-, tri-, tetra-.
  - List substituents alphabetically (ignoring multipliers). Examples follow:
    - $CH_3C(CH_3)_2CH_2CH(CH_3)_2$ is 2,2,4-trimethylpentane

      Counting from the other end would give 2,4,4: 2,2,4 gives the lower number at the first point of difference.

    - $HOCH_2CH_2OH$ is ethane-1,2-diol

      You may have seen this written without the 'e' before the hyphen, but that name is incorrect.

    - $HOCH_2CH_2CH_2CH(CH_3)_2$ is 4-methylpentan-1-ol
    - $CH_3CH_2COCH_2CH_3$ is pentan-3-one
    - $C_6H_5CH_2CH_3$ is ethylbenzene
    - $C_6H_5COCH_3$ is phenylethanone
    - $C_6H_5NHCOCH_3$ is *N*-phenylethanamide

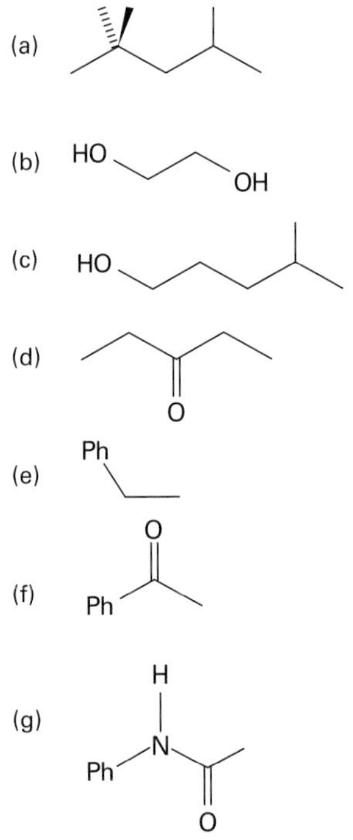

Figure 16.4 shows the skeletal formulae for these seven molecules.

- If there is more than one functional group, we give precedence (for carbon-containing functional groups) to the more oxidized group. So a carboxylic acid group ($-COOH$) takes precedence over a carbonyl group ($-CHO/-COR$) which takes precedence over a hydroxyl group ($-OH$).
  - See Taking the Next Step 16.2 for the link to functional group level.

FIGURE 16.4 The skeletal formulae for (a) 2,2,4-trimethylpentane, (b) ethane-1,2-diol, (c) 4-methylpentan-1-ol, (d) pentan-3-one, (e) ethylbenzene, (f) phenylethanone, and (g) *N*-phenylethanamide. Ph stands for the phenyl group C_6H_5.

- Examples follow:
  - $HOCH_2CH_2CH_2CHO$ is 4-hydroxybutanal
  - $HOCH_2CH_2COOH$ is 3-hydroxypropanoic acid
  - CH_3COCH_2COOH is 3-oxobutanoic acid

Figure 16.5 shows the skeletal formulae for these three molecules.

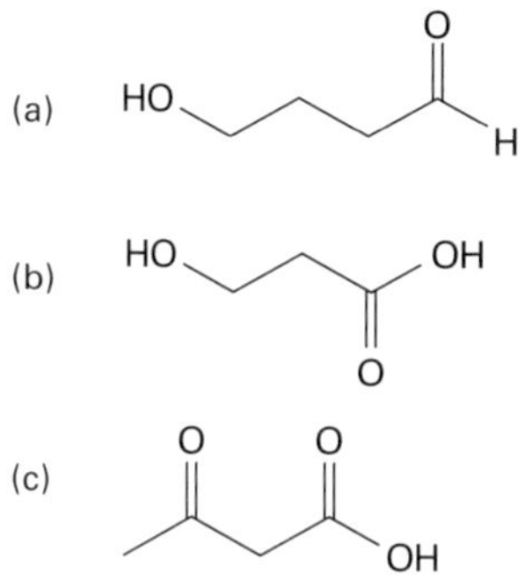

FIGURE 16.5 The skeletal formulae for (a) 4-hydroxybutanal, (b) 3-hydroxypropanoic acid, and (c) 3-oxobutanoic acid.

TAKING THE NEXT STEP 16.2

What does the functional group level tell us?

The **functional group level** (FGL) of a particular carbon atom specifies the number of bonds to atoms that are *more electronegative than carbon*.

The carbon atom singly bonded to oxygen in an alcohol is at functional group level 1 (FGL 1). The carbon atom doubly bonded to oxygen in an aldehyde or a ketone is at FGL 2. The carbon atom doubly bonded to one oxygen atom and singly bonded to another oxygen atom in a carboxylic acid is at FGL 3. The compound CO_2 is at FGL 4.

To turn a carbon atom at a lower FGL to a higher FGL requires an oxidizing agent; to turn a carbon atom at a higher FGL to a lower FGL requires a reducing agent. The carbon atom in a nitrile RCN is at FGL 3, so we can predict that we need a reducing agent to convert it into the amine RCH_2NH_2 (carbon atom at FGL 1). $LiAlH_4$ (Section 22.3) can reduce RCN to RCH_2NH_2. On the other hand, we do not need oxidation or reduction to convert RCN to RCOOH: hydrolysis (reaction with water) under acid catalysis is sufficient.

- At university, you will frequently come across several important molecules that are still called by their traditional names; see Taking the Next Step 16.3.

TAKING THE NEXT STEP 16.3

How can we translate between traditional and IUPAC names?

Traditional	IUPAC	Structure
acetaldehyde	ethanal	CH_3CHO
acetic acid	ethanoic acid	CH_3COOH
acetone	propanone	CH_3COCH_3
aniline	phenylamine	$C_6H_5NH_2$
diethyl ether (ether)	ethoxyethane	$CH_3CH_2OCH_2CH_3$
ethyl acetate	ethyl ethanoate	$CH_3COOCH_2CH_3$
ethylene	ethene	C_2H_4
formic acid	methanoic acid	$HCOOH$
glycerol	propane-1,2,3-triol	$HOCH_2CH(OH)CH_2OH$
lactic acid	2-hydroxypropanoic acid	$CH_3CH(OH)COOH$
tartaric acid	2,3-dihydroxybutanedioic acid	$HOOCCH(OH)CH(OH)COOH$
toluene	methylbenzene	$C_6H_5CH_3$

You might be surprised by the fact that acetone and acetic acid have similar names despite their different numbers of carbon atoms. Vinegar (Latin *acetum*) is a dilute aqueous solution of acetic acid (ethanoic acid). Acetone is the volatile compound made by strongly heating calcium acetate (calcium ethanoate).

16.3 ISOMERISM

- There are two major classes of isomers: structural isomers and stereoisomers.
- **Structural isomers** have the same molecular formula but different structural formulae.
  - We classify structural isomers as chain, position, and functional group isomers.
- **Chain isomers** occur when we can arrange the carbon skeleton in at least two ways (e.g. butane and methylpropane).
  - Chain isomers have very similar chemical properties but different physical properties (branched chain isomers have weaker intermolecular forces and hence lower boiling points).
- **Position isomers** occur when the same functional group can be in at least two places (e.g. 1-bromopropane and 2-bromopropane)
  - Position isomers have slightly different physical properties; they may have similar chemical properties *or* their chemical properties may differ.
    - For example, pentan-2-ol and pentan-3-ol are both secondary alcohols (Section 21.4) and are oxidized to ketones. However, pentan-1-ol is a primary alcohol and is oxidized to an aldehyde or a carboxylic acid.
- **Functional group isomers** have different functional groups and so necessarily have different chemical properties:
  - Propanal (CH_3CH_2CHO) is an aldehyde whereas propanone (CH_3COCH_3) is a ketone.
  - Propanoic acid (CH_3CH_2COOH) is a carboxylic acid whereas methyl ethanoate (CH_3COOCH_3) and ethyl methanoate ($HCOOCH_2CH_3$) are esters.
- **Stereoisomers** have the same structural formula but different arrangements in space.
  - We classify stereoisomers as *E-Z* isomers (obsolescent name geometrical isomers) and optical isomers. (Stereoisomers also occur in inorganic chemistry, Section 15.5.)
    - Both rely on the assignment of priority to different groups. We call the rules for assigning **priority** the **Cahn–Ingold–Prelog (CIP) rules**. Atoms with higher atomic numbers have higher priority. If the first atom encountered in two groups is the same, move down the group comparing each atom in turn. So, for example, −Br has priority over −Cl, which has priority over −COOH, which has priority over −CH_2CH_3, which has priority over −CH_3.

Atoms joined by double bonds need to be counted twice, so the CH_2 in a −CH=CH_2 double bond would count as two carbons and so have priority over −CH_2CH_3.

- **E-Z isomers** occur when there is a carbon–carbon double bond in the molecule. This happens because of *restricted rotation about a double bond* (as the π bond, Figure 2.8, must be broken).

  - If the higher-priority group at one end of the double bond is on the **opposite** side of the bond to the higher-priority group at the other end, as shown in Figure 16.6(a), we label the isomer as **E** (from *entgegen*, German for 'opposite'). If there is one hydrogen atom at each end, we can also use the label **trans** (Latin for 'across').

  - If the higher-priority group at one end of the double bond is on the **same** side of the bond as the higher-priority group at the other end, as shown in Figure 16.6(b), we label the isomer as **Z** (from *zusammen*, German for 'together'; mnemonic: on **z**ee **z**ame **z**ide). If there is one hydrogen atom at each end, we can also use the label **cis** (Latin for 'on this side of').

- Figure 16.7 shows three other examples.

- **Optical isomers** occur when a molecule is chiral: a **chiral** molecule is **not superimposable on its mirror image**. The most common example occurs when a molecule has an **asymmetric carbon atom** (one that has four different groups attached, as shown in Figure 16.8) which is also called a **stereogenic centre** and commonly but less accurately called a **chiral centre**. A Deeper Look 16.1 discusses other ways in which chirality can arise.

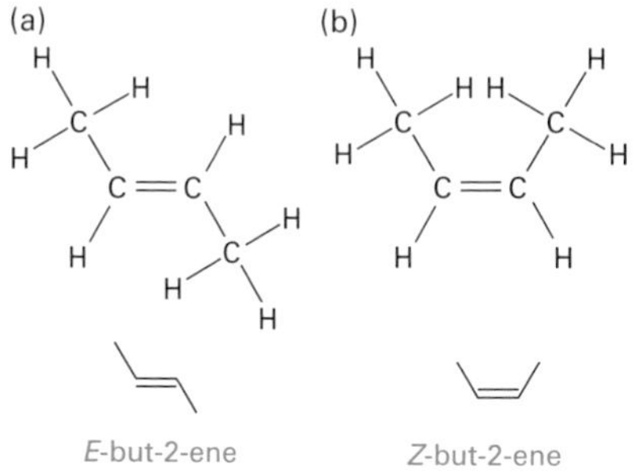

FIGURE 16.6 (a) (*E*)-but-2-ene, and (b) (*Z*)-but-2-ene.

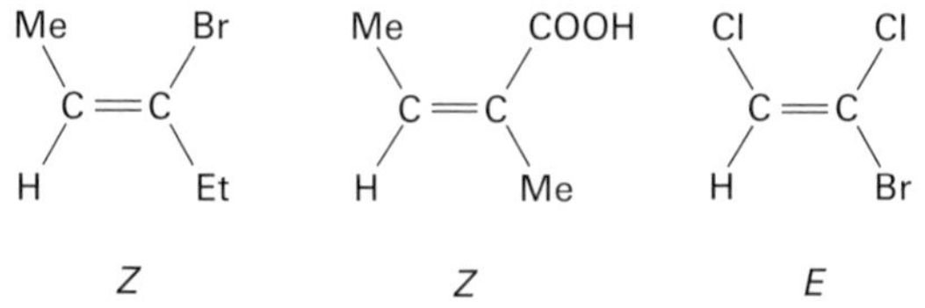

FIGURE 16.7 Examples of *E*/*Z* notation.

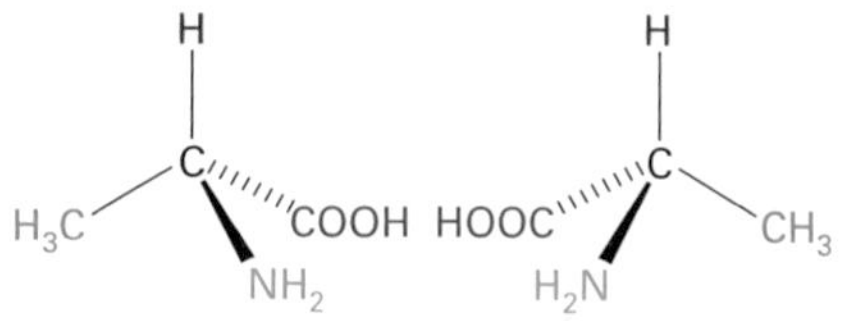

FIGURE 16.8 The two optical isomers of alanine (Table 24.1) are mirror images.

A DEEPER LOOK 16.1

Are there other forms of chirality?

Sometimes it is possible to get chirality from a ring structure. Figure 16.9 shows the chiral molecule Feist's acid, which was first made in the 1890s, but its structure was only solved when NMR (Section 26.3) became widely available in the 1950s. Figure 16.10 shows that it is also possible for chiral structures to occur in **allenes**, which have two double bonds immediately next to each other (in which case the π bonds are then necessarily at right angles to each other).

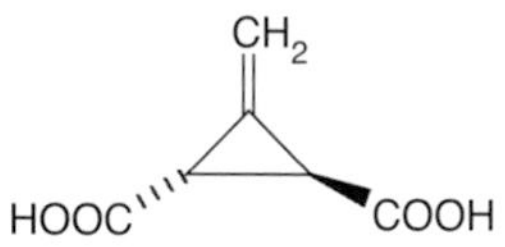

FIGURE 16.9 Feist's acid is chiral.

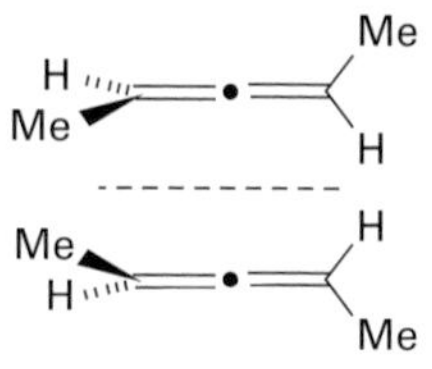

FIGURE 16.10 The two optical isomers of dimethylallene.

The polarimeter measures the direction of rotation (i.e. it specifies the + or − label). It does not identify whether the enantiomer is *R* or *S*, see Taking the Next Step 16.4.

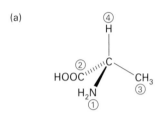

(a)

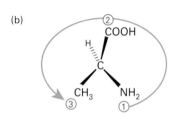

(b)

FIGURE 16.11 (a) The chiral molecule alanine (2-aminopropanoic acid). (b) Draw the molecule with the lowest-priority substituent pointing into the plane of the paper. As the movement from substituent 1 to 2 to 3 is anticlockwise, use the label **S**.

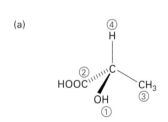

(a)

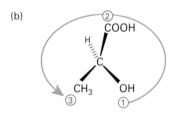

(b)

FIGURE 16.12 (a) The chiral molecule 2-hydroxypropanoic acid (lactic acid). (b) As the movement from substituent 1 to 2 to 3 is anticlockwise, use the label **S**. Lactic acid produced by anaerobic respiration in animal muscles is (*S*)-lactic acid.

- When there is only one asymmetric carbon atom, a pair of **enantiomers** exist which differ because one enantiomer **rotates the plane of plane-polarized light** to the right (**dextrorotatory, +**), whereas the other enantiomer rotates the plane of plane-polarized light to the left (**laevorotatory, −**). We measure the direction (and angle) of rotation using a **polarimeter**.

- Taking the Next Step 16.4 describes an additional label (*R* or *S*, frequently used at university) for the two enantiomers.

- Taking the Next Step 16.5 explains one special case that can occur when two asymmetric carbon atoms are present. (There are many molecules with more than two asymmetric carbon atoms, including sugars such as glucose.)

- The chemical properties of enantiomers are identical except when they are interacting with other chiral molecules. This has particular significance in biochemistry. For example, the (+)-enantiomer of ibuprofen provides pain relief by inhibiting the enzyme cyclooxygenase (COX): the (−)-enantiomer is not a COX inhibitor. The odours of spearmint and caraway seed are caused by the two enantiomers of carvone.

TAKING THE NEXT STEP 16.4

What do the labels *R* and *S* mean?

We can use the CIP rules to label the two different enantiomers in a pair. For a molecule such as the amino acid alanine $H_2NCH(CH_3)COOH$, assign each substituent a priority. Figure 16.11 shows how to draw the molecule with the lowest-priority substituent (H in this case) pointing into the plane of the paper.

    If the movement from substituent 1 to 2 to 3 is clockwise, use the label **R** (Latin *rectus*). (If you turn a steering wheel clockwise, you would be turning right.) If the movement from substituent 1 to 2 to 3 is anticlockwise, use the label **S** (Latin *sinister*). So the enantiomer drawn in Figure 16.11 is (*S*)-alanine, the one found in nature. See also Figure 16.12.

> An −OH group takes priority over a −COOH group using the CIP rules. However, in IUPAC nomenclature the molecule $CH_3CH(OH)COOH$ is called 2-hydroxypropanoic acid. The CIP rules are used to define the *stereochemistry* whereas the IUPAC rules are used to *name* the molecule, so the rules are not in conflict with each other.

TAKING THE NEXT STEP 16.5

What is a *meso* isomer?

When a molecule has two asymmetric carbon atoms that have the **same** atoms or groups attached to them, only three isomers (rather than the expected $2^2 = 4$) exist. One isomer, the **meso** isomer, as drawn in Figure 16.13, does not show optical activity. The key feature that makes the *meso* isomer not optically active is that it has a plane of symmetry. The rotation caused by one end of the molecule cancels the equal but opposite rotation caused by the other end. One example occurs in tartaric acid (2,3-dihydroxy-butanedioic acid), a common ingredient in fizzy drinks.

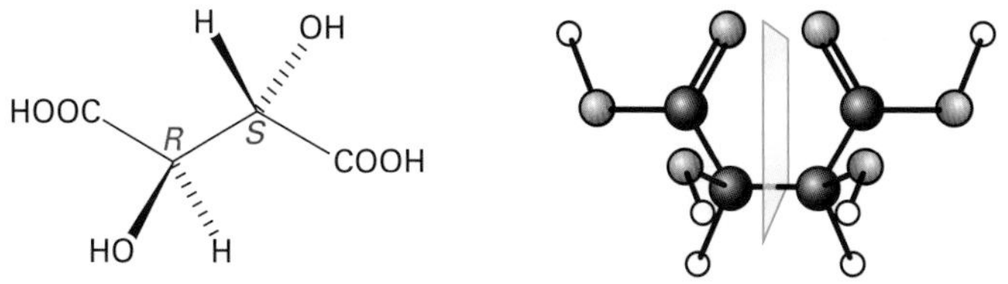

FIGURE 16.13 The *meso* isomer of tartaric acid. The plane of symmetry ensures that the *meso* isomer does not show optical activity. See Taking the Next Step 16.4 for an explanation of the labels *R* and *S*.

- A **racemic mixture** is an **equimolar mixture of the two enantiomers**, which therefore does not rotate the plane of plane-polarized light.
  - When hydrogen cyanide adds to an aldehyde or an unsymmetrical ketone (Section 22.4), the fact that the *carbonyl group is planar* (as shown in Figure 16.14) means that attack can occur *equally easily* from above or below the plane, resulting in a racemic mixture.

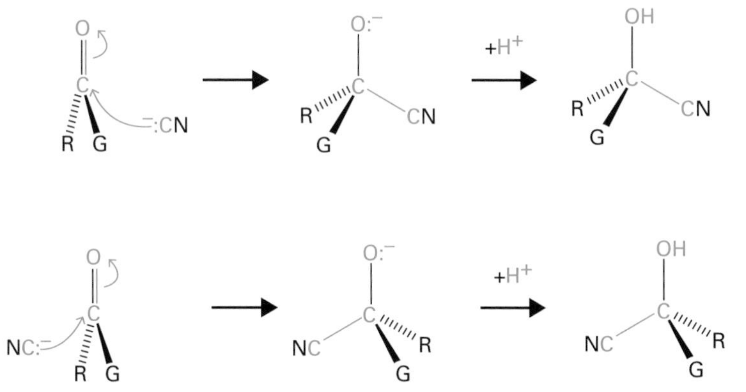

FIGURE 16.14 The cyanide ion can attack from above or below the plane equally easily, so a racemic mixture forms. G can be H or R'.

Taking the Next Step 16.6 describes how we can separate a racemic mixture—a process called **resolution**.

A Deeper Look 16.2 describes how it is also possible to produce more of one of the two enantiomers. This happens very commonly in biological systems.

TAKING THE NEXT STEP 16.6

How can we resolve a racemic mixture?

Louis Pasteur achieved the first successful **resolution** (separation into the two enantiomers) of a racemic mixture in 1848 when he noticed under the microscope that there were two different-shaped crystals of sodium ammonium tartrate (Taking the Next Step 16.3), as illustrated in Figure 16.15. This allowed him to separate them by simply picking out the two different shapes. They turned out to be the (+)-enantiomer and the (−)-enantiomer.

FIGURE 16.15 The crystals of sodium ammonium tartrate have different shapes under the microscope.

More usually, more sophisticated methods are needed. The following technique was also discovered by Pasteur. We can resolve optically active acids by reaction with an optically active base such as quinine. The resulting salts are **diastereoisomers** (stereoisomers that are not enantiomers), which have sufficiently different physical properties that they can be separated, often by chromatography.

A DEEPER LOOK 16.2

How can we achieve asymmetric synthesis?

One way to achieve more of one enantiomer (an **enantiomeric excess**) is to start from an already chiral species (**chiral pool synthesis**), such as a natural amino acid. If you do not do any reaction that affects the asymmetric carbon atom, it is reasonable to assume that its chirality remains unchanged.

A second way involves using a chiral catalyst (**asymmetric catalysis**). The first example was found in 1968 by William Knowles, who used a chiral rhodium complex as a homogeneous catalyst (Section 14.4) for hydrogenation to produce L-DOPA, used in the clinical treatment of Parkinson's disease. A chiral ruthenium complex introduced by Ryoji Noyori is used industrially to make the non-steroidal anti-inflammatory drug naproxen.

16.4 ORGANIC REACTIONS

- We classify reactive species in organic reactions as radicals, nucleophiles, or electrophiles.
  - A **radical** has a single, **unpaired electron**. Radicals are typically highly reactive.
  - A **nucleophile** is an **electron-pair donor**. A nucleophile is typically a negative ion or a neutral molecule with lone pairs.

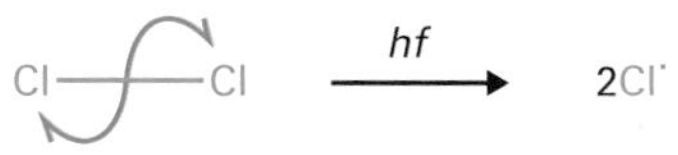

FIGURE 16.16 Homolytic fission of a chlorine molecule into two chlorine radicals. Section 17.3 explains the 'fish-hooks'.

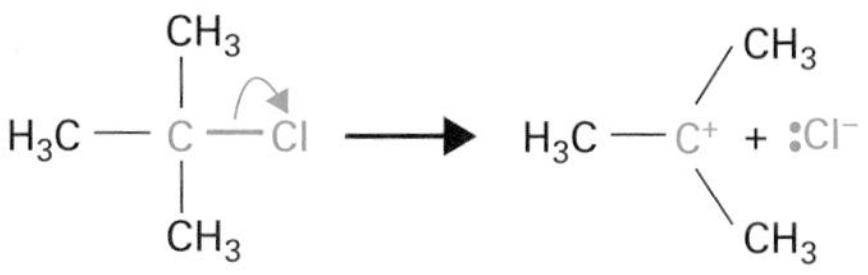

FIGURE 16.17 Heterolytic fission produces a positive ion and a negative ion. The blue colour highlights the parts of any species that are changing during the reaction.

- An **electrophile** is an **electron-pair acceptor**. The simplest electrophile is the hydrogen ion, H^+.
- Bonds can break in one of two ways:
  - **Homolytic fission** produces two radicals, as shown in Figure 16.16. This is often caused by the presence of UV light (symbolized by *hf*).
  - **Heterolytic fission** produces a positive ion and a negative ion, as shown in Figure 16.17.
- **Curly arrows** show the **movement of an electron pair**.
- We can classify organic reactions as substitution, addition, elimination, condensation, polymerization, and redox reactions.
- In a **substitution** reaction, one part of a molecule is replaced by a different group. One example is the nucleophilic substitution reaction of CH_3CH_2Br with NaOH:

$$CH_3CH_2Br + NaOH \rightarrow CH_3CH_2OH + NaBr$$

- In an **addition** reaction, two molecules become one. One example is the electrophilic addition reaction of C_2H_4 with HBr:

$$H_2C = CH_2 + HBr \rightarrow CH_3CH_2Br$$

- In an **elimination** reaction, one molecule becomes two. One example is the elimination reaction of $(CH_3)_3CBr$ with hot concentrated alcoholic KOH:

$$(CH_3)_3CBr + KOH \rightarrow (CH_3)_2C = CH_2 + KBr + H_2O$$

- In a **condensation (addition–elimination)** reaction, an initial addition reaction is followed by an elimination reaction. One example is the condensation reaction of CH_3COCl with CH_3CH_2OH:

$$CH_3COCl + CH_3CH_2OH \rightarrow CH_3COOCH_2CH_3 + HCl$$

- In a **polymerization** reaction, many small monomer molecules combine to form one large polymer molecule. The reaction occurs via addition or condensation (Section 25.1).
- In a **redox** reaction, the organic molecule is either oxidized or reduced.
  - A typical reagent for oxidation is acidified potassium dichromate(VI), which can, for example, oxidize secondary alcohols to ketones. (The dichromate(VI) anion is reduced to the chromium(III) cation and this colour change from orange to green, Section 21.4, is a useful test.)

Electrophiles are neutral or positively charged species with an empty atomic orbital (or a low-energy antibonding orbital) that can accept an electron pair.

16

— One reagent for reduction is sodium tetrahydridoborate (sodium borohydride), which can, for example, reduce ketones to secondary alcohols, Section 22.3.

A Deeper Look 16.3 introduces an idea that is very useful at university.

A DEEPER LOOK 16.3

What do HOMO and LUMO mean?

At university, you will find that many reactions can be best explained by considering the interaction between the highest occupied molecular orbital (**HOMO**) of the *nucleophile* and the lowest unoccupied molecular orbital (**LUMO**) of the *electrophile*. We call the HOMO and the LUMO the **frontier orbitals**. See Figure 19.15, A Deeper Look 20.1, and A Deeper Look 22.4.

HYDROCARBONS: ALKANES

17.1 THE OIL INDUSTRY: FRACTIONAL DISTILLATION

- A **hydrocarbon** contains the elements hydrogen and carbon only.

- The two most important environmental sources of hydrocarbons are **natural gas** (which is mostly methane) and crude oil.

- **Crude oil (petroleum)** is a complex mixture of hydrocarbons, mostly alkanes, formed from the compressed remains of sea creatures; hence crude oil's composition varies from place to place.

- **Fractional distillation** allows separation of the mixture and relies on the different **fractions** having different boiling points, because intermolecular forces increase as the chain length of the molecule increases (Section 4.2). Taking the Next Step 17.1 explains the fractional distillation process. (It is not always possible to separate liquid mixtures completely by fractional distillation, see A Deeper Look 17.1.)

TAKING THE NEXT STEP 17.1

How does fractional distillation work?

Raoult's law often adequately describes the vapour pressure above a mixture of two liquids. **Raoult's law** states that the contribution of each liquid to the total vapour pressure p is proportional to its mole fraction (Section 6.1) in the liquid; this produces a linear graph of vapour pressure as a function of composition, as shown in Figure 17.1.

The temperature at which the total vapour pressure reaches the external pressure (and hence boiling takes place) produces a boiling-point graph which is *not* linear, as shown for a mixture of methanol and water in Figure 17.2.

The vapour produced when the liquid mixture boils is richer in the more volatile component, and this allows separation of the two components, as shown in Figure 17.3.

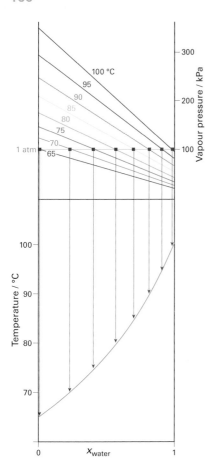

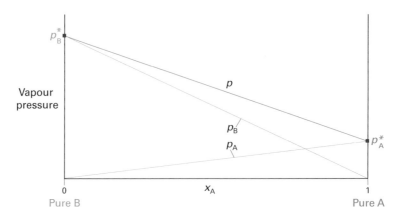

FIGURE 17.1 Raoult's law states that the vapour pressure of liquid A, p_A, in a mixture is proportional to its mole fraction, x_A, in the liquid. The vapour pressure of pure liquid A is $p*_A$.

FIGURE 17.2 The vapour pressure for a mixture of methanol and water. As the vapour pressure of the more volatile component (here methanol) rises faster as temperature increases, the boiling point line is curved.

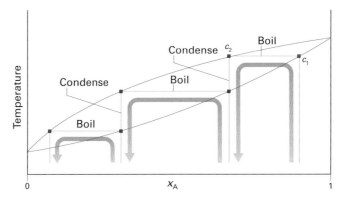

FIGURE 17.3 The vapour above the boiling liquid mixture is richer in the more volatile component: here A is the less volatile component. The composition c_1 of the boiling liquid mixture produces a vapour composition c_2 with less A present. This allows separation of the two components after a number of boiling/condensing steps.

A DEEPER LOOK 17.1

Azeotropes

We cannot completely separate some liquid mixtures, such as a mixture of ethanol and water, by fractional distillation: fractional distillation of ethanol and water can only produce 96% ethanol. We call this particular mixture an **azeotrope**. The azeotrope behaves like a pure substance in that, when distilled, the composition of the vapour is identical to that of the liquid. Hence no further separation occurs, as is shown in Figure 17.4.

We can take advantage of azeotrope formation in reactions that require the removal of a water impurity. We often carry out such reactions in refluxing toluene (methylbenzene) under what is called a **Dean–Stark head**, as illustrated in Figure 17.5. The azeotrope when condensed separates out into two layers, as toluene is immiscible with water: we can then return the upper toluene layer to the reaction vessel.

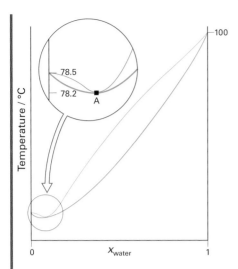

FIGURE 17.4 Ethanol and water form an azeotrope (A), which we cannot separate further by fractional distillation.

FIGURE 17.5 A Dean–Stark head.

17.2 ALKANES

- **Alkanes** are saturated hydrocarbons, whose general formula is C_nH_{2n+2}. In a **saturated** hydrocarbon, each carbon atom is linked to **four** other carbon or hydrogen atoms by **single** bonds. The simplest alkane is methane, CH_4: Taking the Next Step 17.2 discusses the bonding in methane. The bonding in other alkanes and in alkyl groups is very similar.

- Alkanes are relatively unreactive because they have strong C–C and C–H bonds (the latter having low polarity). Since the molecules have no significant polarity, very harsh conditions are needed to get any reaction, such as setting fire to them (combustion) or creating highly reactive radicals (Section 16.4) which can then react with the alkanes. For example, we can replace a C–H bond with a C–Cl bond using a chlorine radical. We discuss **radical substitution** in Section 17.3.

The industrial process of **thermal cracking** studied pre-university is also a radical process: the breaking of carbon–carbon bonds typically gives two product molecules.

17

TAKING THE NEXT STEP 17.2

What is the bonding in methane?

Methane, the simplest alkane, consists of a central carbon atom with four hydrogen atoms bonded covalently: the carbon atom has no lone pairs as all its electrons are used in bonding. VSEPR theory (Section 2.4) predicts that, with no lone pairs and four bond pairs, methane's shape is tetrahedral. We can predict that the bond angle is the perfect tetrahedral angle (109.5°), because the four bond pairs repel each other equally. This is confirmed experimentally. Carbon is in Group 14 (Figure 11.1) of the periodic table, and the only orbitals carbon has in its valence shell are the 2s and 2p orbitals.

The carbon 2s orbital can overlap with all four of the hydrogen 1s orbitals, with all orbitals in phase (Section 2.1), as shown in Figure 17.6, to

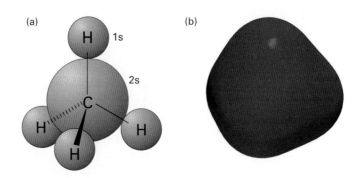

FIGURE 17.6 (a) The carbon 2s orbital can form a bonding molecular orbital by in-phase overlap with the 1s orbitals of all four hydrogen atoms. (b) The resulting molecular orbital is delocalized over all five atoms.

form a delocalized molecular orbital (MO). The electron density in the MO formed from the 2s is symmetrical.

The three carbon 2p orbitals can also interact, as shown in Figure 17.7, but require different phases (Section 2.1) for the hydrogen 1s orbitals in each case. *Summing* the electron density in these three MOs shows that each carbon–hydrogen bond has the same electron density.

Each of the four molecular orbitals is delocalized (Section 2.3). The MO derived from the 2s orbital has no nodal plane (Section 2.1) and is lower in energy than the other three orbitals. This description is supported by experimental measurements and detailed computer calculations; see Figure 17.7.

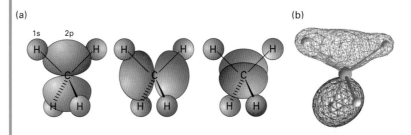

FIGURE 17.7 (a) Each carbon 2p orbital can form a bonding molecular orbital by in-phase overlap with the 1s orbitals of the four hydrogen atoms. (Each p orbital requires a different set of phases for the hydrogen 1s orbitals.) These three orbitals are all of the same energy (are degenerate, Section 1.4). (b) One of the computed MOs derived from the 2p orbitals.

You will frequently encounter an alternative simpler picture at university. **Hybridization** maintains the key idea of the symmetric electron density in each C–H bond; see Taking the Next Step 17.3. A description of the bonding made using hybrid orbitals is a rough but nevertheless useful guide to what is going on.

TAKING THE NEXT STEP 17.3

What is the sp³ hybridization model for alkanes?

In the hybridization model of methane, we construct four equivalent bonds by first combining together the 2s and **all three** 2p atomic orbitals on carbon to form four equivalent **hybrid** orbitals. We call each one an **sp³** (pronounced s-p-three not s-p-cubed) hybrid orbital. Each sp³ hybrid orbital has a major lobe and a minor lobe, as shown in Figure 17.8. The number of hybrid orbitals formed is always the same as the number of starting atomic orbitals.

The four sp³ hybrid orbitals point towards the corners of a tetrahedron, as shown in Figure 17.9. We can form each C–H bond in methane by overlapping one sp³ hybrid orbital with the 1s orbital of one hydrogen atom. Each C–H bond in *ethane* is also formed by overlapping a carbon sp³ hybrid orbital with the 1s orbital of a hydrogen atom, while the C–C bond is formed by overlapping two sp³ hybrids, one from each carbon atom.

Any tetrahedral carbon atom forming bonds to four atoms will be described as being sp³ hybridized.

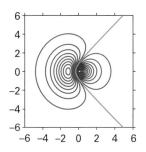

FIGURE 17.8 An sp³ hybrid orbital. The scale is in Bohr radii (Figure 1.7). The green lines indicate nodal planes (Section 2.1).

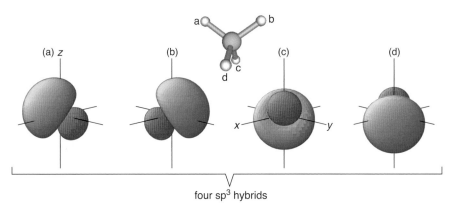

FIGURE 17.9 The four sp³ hybrids point towards the corners of a tetrahedron.

17.3 RADICAL SUBSTITUTION OF ALKANES

- Because they are saturated (Section 17.2) hydrocarbons, every carbon atom forms four bonds and so cannot form any additional bonds. Therefore alkanes can only react by *substitution*. Furthermore, because the molecule has no significant polarity, the only attacking species that can react is a *radical*: radicals are typically highly reactive (Section 16.4). A radical has a single, unpaired electron, indicated by a **dot** in mechanisms.

- The most important reaction of alkanes (apart from combustion) is halogenation. **Photochemical halogenation** of alkanes occurs by reaction with **a halogen** in the presence of **ultraviolet (UV) light**. As an example, we will discuss chlorination (but bromination occurs in exactly the same way).

Fluorination would be dangerously violent; iodination is unsuccessful as the reaction is endothermic.

The symbol for frequency used at university by chemists is usually the Greek letter nu, ν (A Deeper Look 1.3).

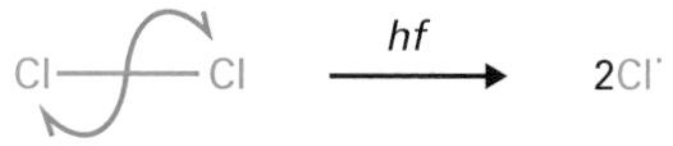

$$Cl \overset{hf}{\longrightarrow} 2Cl^{\cdot}$$

FIGURE 17.10 Chain initiation.

We can identify radicals using a spectroscopic technique called **electron spin resonance** (**ESR**), which is analogous to NMR (Section 26.3).

– $CH_4 + Cl_2 \rightarrow CH_3Cl + HCl$

- The mechanism is a **radical chain reaction**.
- The first stage is **chain initiation**.
  – The UV light (indicated by *hf* in Figure 17.10) causes **homolytic fission** (Section 16.4, where the chlorine–chlorine bond breaks such that the two atoms involved each retain one electron) to form **chlorine radicals**, as shown in Figure 17.10. Notice how this figure makes use of **fish-hooks.**

    A *single*-headed arrow, usually called a **fish-hook**, indicates the movement of a *single* electron. It is very important not to confuse these fish-hooks with the very familiar *double*-headed curly arrows (Section 16.4), which show the motion of an *electron pair*.

- The second stage is **chain propagation**.
  – In this *pair* of steps, the very reactive radicals that are generated react with other molecules to produce a new radical that can then **continue the chain reaction**. In Figure 17.11, the initial chlorine radical **abstracts** (removes) a hydrogen atom from methane to make a **methyl radical** and hydrogen chloride. This methyl radical can then react with a chlorine molecule to make chloromethane, *regenerating* a new chlorine radical.

- The final stage is **chain termination**.
  – As the radicals are so reactive, they can react with each other to make a stable molecule. This stops the chain reaction from continuing, hence why we call such a reaction 'chain termination'. Figure 17.12 shows three examples. The combination of two methyl radicals produces ethane C_2H_6 (providing good evidence for the mechanism).

- This is a very poor reaction synthetically as the radicals are highly reactive and **further substitution** is possible, giving a range of products including CH_2Cl_2, $CHCl_3$, and CCl_4 in the case of methane.

- Radical substitution of the side chains of aromatic compounds is *much* more selective; see A Deeper Look 17.2.

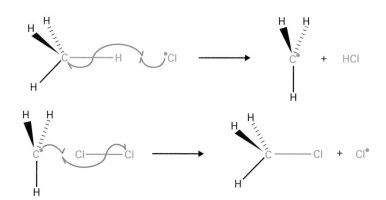

FIGURE 17.11 Chain propagation.

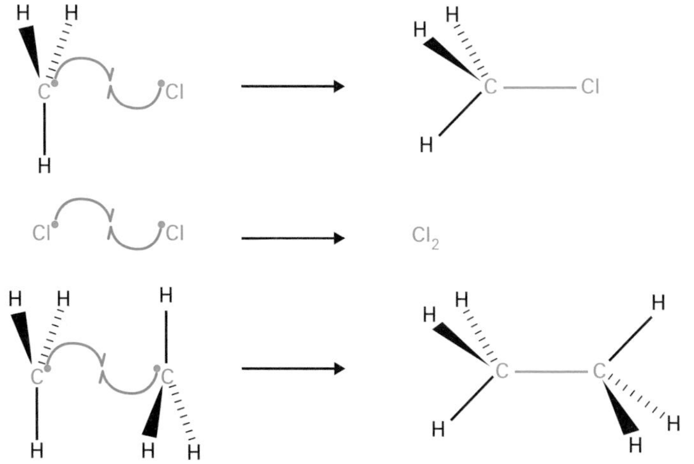

FIGURE 17.12 Chain termination.

A DEEPER LOOK 17.2

Radical substitution of the side chains of arenes

Chlorine radicals normally attack alkanes fairly indiscriminately. However, in an alkylbenzene, radical attack occurs preferentially at the carbon atom *nearest* the ring because the radical formed during the reaction is delocalized with the benzene ring (Section 19.1), as shown in Figure 17.13. The dominant substitution product is $C_6H_5CHClCH_3$.

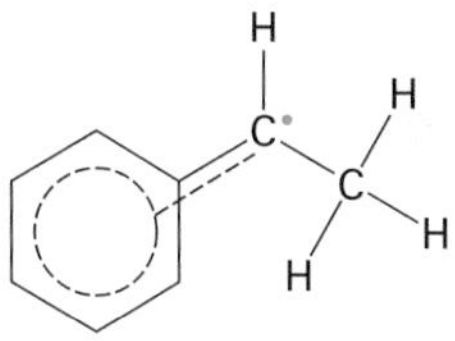

FIGURE 17.13 This particular radical formed from ethylbenzene is stabilized by delocalization with the benzene ring.

17.4 COMBUSTION OF ALKANES

- Combustion of alkanes is their most important reaction commercially. Much of our transportation, central heating, and cooking currently relies on the combustion of petrol, diesel, or natural gas.

- Combustion is also a radical chain reaction.

- **Complete combustion** produces carbon dioxide and water, so for octane:

  $$- 2C_8H_{18} + 25O_2 \rightarrow 16CO_2 + 18H_2O$$

  Incomplete combustion also produces carbon monoxide (a particular danger if central heating systems are poorly ventilated) and sometimes carbon (a particular danger from the combustion of diesel).

A petrol with an **octane number** of 95 burns as efficiently as a mixture of 95% 2,2,4-trimethylpentane (Section 16.2), which is the isomer of octane traditionally called 'isooctane' by petroleum engineers, plus 5% heptane.

17

HYDROCARBONS: ALKENES

18.1 INTRODUCTION TO THE ALKENES

At university you may come across alkenes called **olefins**. Ethene may also be called **ethylene**.

- Alkenes are unsaturated hydrocarbons, whose general formula is C_nH_{2n}. An **unsaturated** hydrocarbon contains a carbon–carbon multiple bond, in this case a double bond. The double bond consists of a **σ bond** and a **π bond** (Section 2.1).

- The simplest alkene **ethene**, C_2H_4, is a planar molecule with an HCH bond angle of 117°, just smaller than the expected angle of 120° for a trigonal planar structure (Section 2.4) because the double bond is a little more repulsive than a single bond. Taking the Next Step 18.1 introduces sp^2 hybrid orbitals. The next alkene, C_3H_6, is **propene**.

> ### TAKING THE NEXT STEP 18.1
>
> ### What hybridization does an alkene show?
>
> A trigonal planar carbon atom, as for example in ethene, requires a different type of hybridization (Section 17.2) from that in alkanes. The s orbital and **two** of the p orbitals are hybridized to form three **sp^2** (pronounced s-p-two) hybrid orbitals, which point towards the corners of an equilateral triangle, as shown in Figure 18.1.

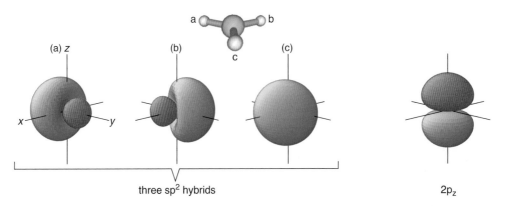

FIGURE 18.1 The three sp^2 hybrid orbitals in ethene lie in a plane and point towards the corners of an equilateral triangle, giving a trigonal planar geometry. One perpendicular p orbital remains unhybridized.

We can form each C–H bond in ethene by overlapping a carbon sp² hybrid orbital with the 1s orbital of a hydrogen atom. By contrast, we can form the C–C σ bond by overlapping two sp² hybrid orbitals, one from each carbon atom. This leaves one p orbital on each carbon atom: they overlap sideways (Section 2.1) to form the π bond, as shown in Figure 18.2.

Any trigonal planar carbon atom forming bonds to three atoms will be described at university as being sp² hybridized.

- Alkenes that have a spine longer than three carbon atoms can show isomerism. For example, butene (C_4H_8) has two different regions where the double bond can be, and can therefore exist as two position isomers (Section 16.3) but-1-ene and but-2-ene, as shown in Figure 18.3.
  - But-1-ene has the double bond joining carbon atoms 1 and 2 whereas but-2-ene has the double bond joining carbon atoms 2 and 3.
  - The latter can have two different spatial arrangements that are not readily interconvertible: the *E/trans* and *Z/cis* isomers. Figure 18.3(b) shows these isomers.
  - *E/trans* and *Z/cis* isomers (Section 16.3) arise because the **double bond resists rotation**: to interconvert the two isomers we need to break the π bond, as shown in Figure 18.4.

- The ethene molecule has a sigma bond and a pi bond between the two carbon atoms, as shown in Figure 18.5. (The sigma bond is considered to arise from overlap of two p orbitals rather than two sp² hybrid orbitals.)

- Ethene is **planar** with **relatively high electron density** between the two carbon atoms, as shown in Figure 18.6.

- In ethene and alkenes with electron-donating substituents (Section 18.3) such as alkyl groups around the double bond, the electron density is relatively high around the carbon–carbon double bond. This high electron density makes the alkene a nucleophile; the characteristic reactions of these types of alkenes are **electrophilic addition reactions**.

FIGURE 18.2 The two remaining perpendicular p orbitals in ethene overlap sideways (Section 2.1) to form the π bond.

Alkynes have the general formula C_nH_{2n-2}: the simplest is ethyne (acetylene), C_2H_2. Alkynes are linear in shape; any carbon atom forming two bonds at 180° to each other will be described at university as being **sp** hybridized.

See Section 14.5 for the mechanism of catalytic hydrogenation of alkenes.

18

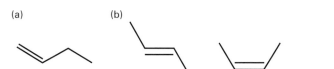

FIGURE 18.3 The isomers of butene: (a) but-1-ene and (b) (*E*)- and (*Z*)-but-2-ene.

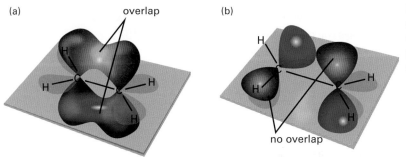

FIGURE 18.4 (a) Rotating one end of the ethene molecule relative to the other reduces the overlap between the p orbitals, until (b) at 90° there is no overlap at all: the π bond is broken.

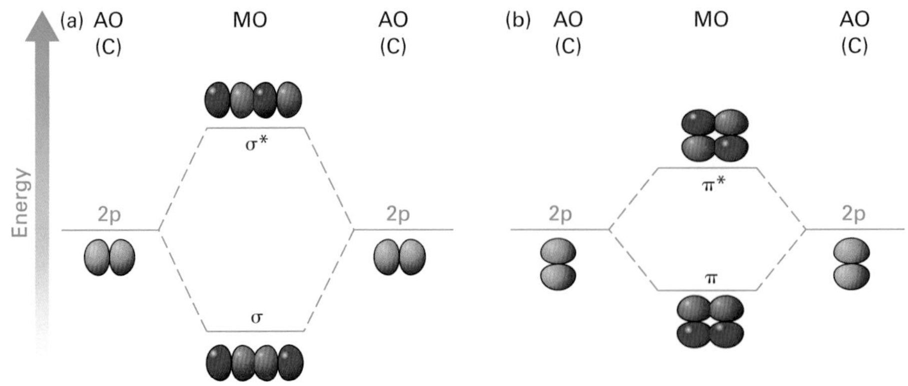

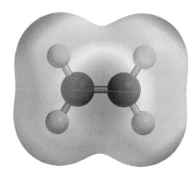

FIGURE 18.5 The formation of bonding (and antibonding) molecular orbitals in ethene: (a) the σ bond, formed by end-on overlap, and (b) the π bond, formed by sideways overlap. The σ* and π* orbitals are not occupied but are important at university for understanding ultraviolet spectra.

FIGURE 18.6 The electrostatic potential map (Section 2.2) for ethene. The red colour indicates the high electron density in the double bond. The Spartan software used to generate this image intentionally does not show multiple bonding, only which atoms are bonded together.

18.2 ALKENES UNDERGO ELECTROPHILIC ADDITION REACTIONS

- Typical electrophilic addition reactions include the addition of hydrogen halides (either in concentrated aqueous solution or in the gas phase), acidified water (typically using sulfuric or phosphoric acids), and halogens.

Electrophilic addition under acidic conditions

- As the proton (hydrogen ion) present in an acidic solution is positively charged, it will be the electrophile in this reaction and is therefore attacked by the π bond of the alkene, Figure 18.7.

- The intermediate that is formed is a **carbocation**, a species in which carbon is carrying a positive charge and forming three bonds. The empty p orbital on the positively charged carbon (see Figure 18.12) makes it an electrophile, leaving it open to attack by nucleophiles. See Taking the Next Step 18.2.

- The nucleophile may be negatively charged, as for a halide ion such as Br⁻ from hydrobromic acid HBr. Attack by the nucleophile could be from above or below the plane around the positively charged carbon, as shown in Figure 18.8.

- For ethene and HBr, for example, the product of the reaction is CH_3CH_2Br regardless of the direction of attack. However, this is not always the case, as we see in A Deeper Look 18.1.

FIGURE 18.7 The high electron density in the π bond attracts the proton, which reacts to form a carbocation.

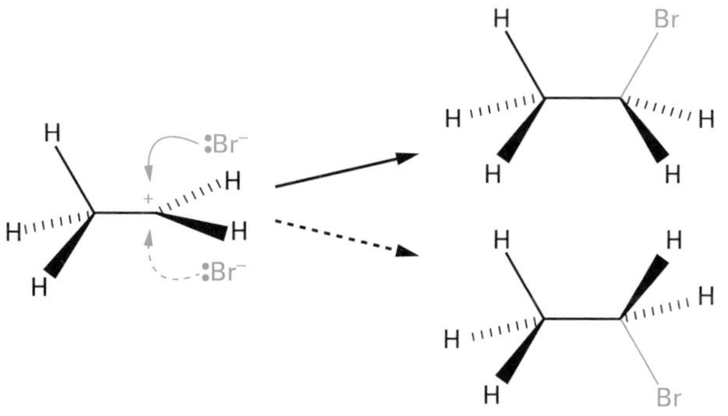

FIGURE 18.8 Attack of the bromide ion could be from above or below the plane around the positively charged carbon.

TAKING THE NEXT STEP 18.2

What hybridization does a carbocation show?

The ethyl carbocation $CH_3CH_2^+$ has two carbon atoms with different hybridization:

- The carbon atom forming four bonds is tetrahedral and sp^3 hybridized.
- The positively charged carbon is trigonal planar and sp^2 hybridized. This leaves one vacant p orbital perpendicular to the plane (see Figure 18.12).

A DEEPER LOOK 18.1

Does the direction of attack matter?

As the carbocation is planar around the positively charged carbon, the nucleophile can attack from either above or below the plane with equal probability, which may result in the formation of a racemic mixture (Section 16.3), as is the case for (E)-but-2-ene, as shown in Figure 18.9.

FIGURE 18.9 Attack of the bromide ion could be from above or below the plane around the positively charged carbon. For (E)-but-2-ene this results in a racemic mixture.

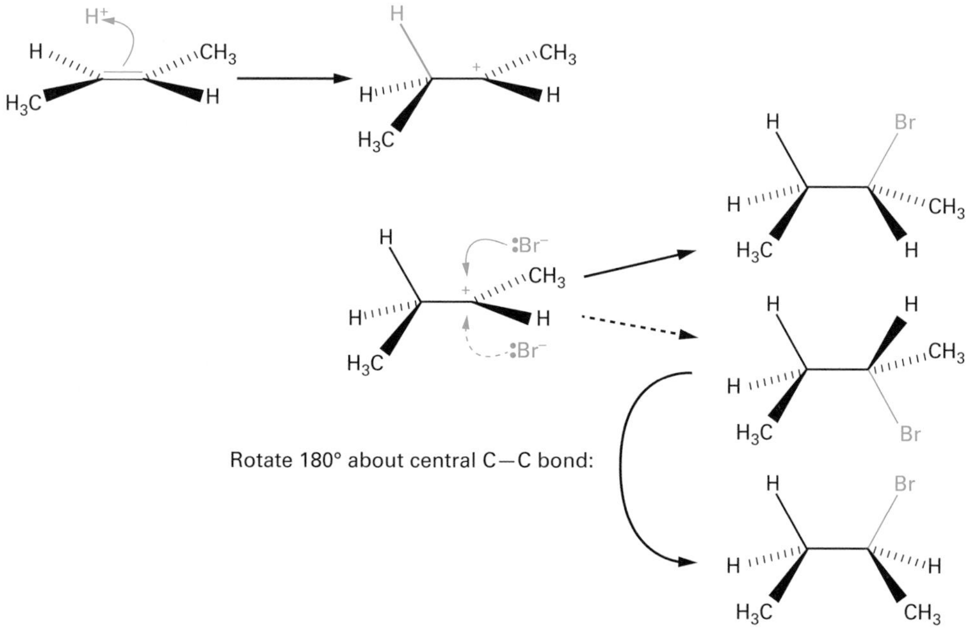

Rotate 180° about central C—C bond:

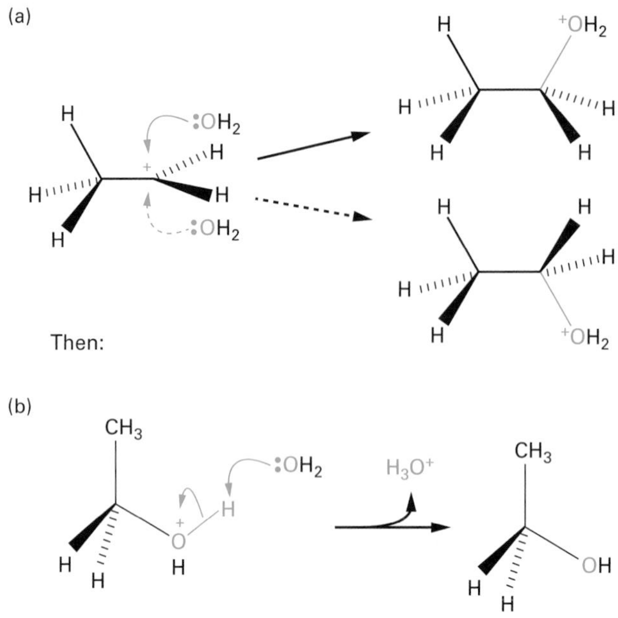

(a)

Then:

(b)

FIGURE 18.10 (a) Attack of a water molecule on the carbocation forms a protonated alcohol. (b) Deprotonation occurs in a subsequent step.

The **direct hydration of ethene** is an important industrial process, as discussed in detail in Section 21.1.

- Any uncharged nucleophile, such as water, will follow exactly the same mechanism as for a negatively charged nucleophile when attacking the carbocation, except that the formation of a stable product requires one further step, most importantly deprotonation in the case of water, Figure 18.10(b). The product formed from ethene is ethanol.

18.3 MARKOVNIKOV'S RULE

- When the alkene is **unsymmetrical** (such as propene, $CH_3CH=CH_2$) and the adduct is also **unsymmetrical** (such as HBr) two isomers could form. However, normally one of the two isomers dominates. In the case of $CH_3CH=CH_2$ and HBr the dominant product is $CH_3CHBrCH_3$. We call this **Markovnikov's rule**, Figure 18.11.

> Vladimir Markovnikov originally stated (in a paper submitted in 1869) that 'when an unsymmetrical alkene bonds with a hydrohalic acid [such as HBr] the halide adds to the carbon atom with *fewer* hydrogen atoms (that is to say, to the carbon atom that is more under the influence of other carbons)'. It is *much* more important to understand the reason behind the rule, as explained below.

A **tertiary carbocation** has *three* alkyl groups attached to the positively charged carbon, a **secondary carbocation** has *two*, and a **primary carbocation** has *one*.

- The idea that the **more stable carbocation forms** explains Markovnikov's rule. A secondary carbocation (see margin) is more stable than a primary carbocation because **alkyl groups are electron-donating**; see Taking the Next Step 18.3. (A tertiary carbocation is even more stable.)

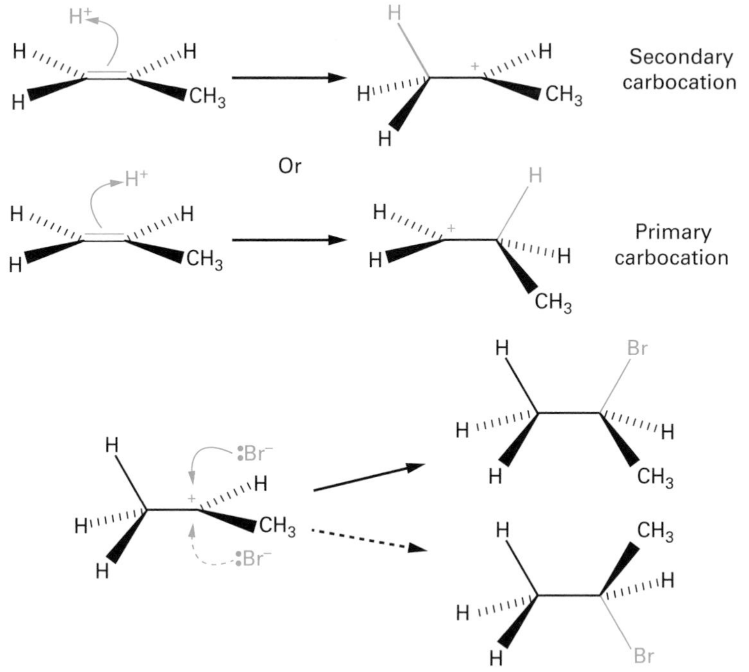

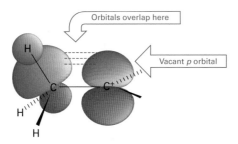

FIGURE 18.11 Protonation of propene could produce two possible carbocations. The secondary carbocation (at the top) is more stable than the primary carbocation, so it is the secondary carbocation which goes on to react with bromide ion.

TAKING THE NEXT STEP 18.3

Why are alkyl groups electron-donating (electron-releasing)?

Figure 18.12 shows that the electron pair in a σ bond of the methyl group (or a general alkyl group) is attracted towards the vacant p orbital of the carbocation. When the electron density shifts in this way, the carbon becomes somewhat less positive. We call this stabilization of a vacant p orbital by delocalization of a σ bond **σ conjugation** or **hyperconjugation**.

FIGURE 18.12 Overlap of the C–H bond of the methyl group in this particular orientation with the vacant p orbital donates electron density into the p orbital.

As the number of alkyl groups attached to the positively charged carbon increases, the electron density attracted to the positively charged carbon increases, reducing the localized positive charge on the carbon. Figure 18.13 illustrates this trend.

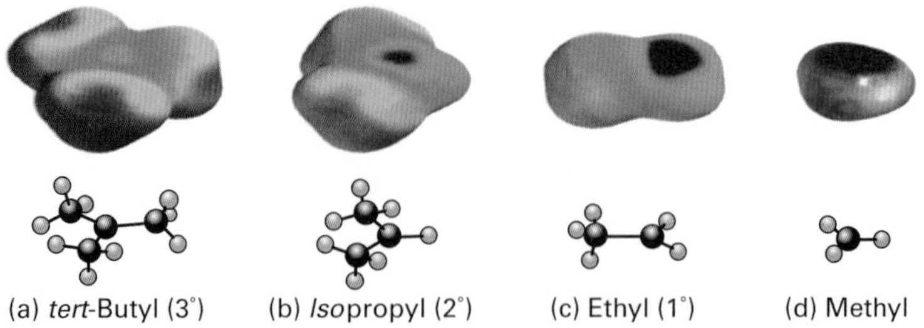

(a) *tert*-Butyl (3°)
(b) *Iso*propyl (2°)
(c) Ethyl (1°)
(d) Methyl

FIGURE 18.13 The electrostatic potential maps (Section 2.2) show the trend from greatest to least stabilization of the positive charge: a more intense blue colour shows a more localized positive charge.

The electrostatic potential maps in Figure 18.13 show the trend from greatest to least stabilization of the positive charge: a more intense blue colour shows a more localized positive charge, which makes the carbocation less stable.

18.4 ADDITION OF A NEUTRAL MOLECULE TO ETHENE

- The mechanism for this addition, shown in Figure 18.14, is very similar to that for addition under acidic conditions.
- The added molecule can be either a molecule with a permanent dipole, such as HBr, or even a non-polar molecule, such as a halogen like Br_2, because the double bond can induce a dipole in Br_2 as it approaches.

Y
X
H
H
H
H
X
H
H
H
H
+ :Y⁻
X
:Y⁻
H
H
H
H
:Y⁻
X
Y
H
H
H
H
X
H
H
H
H
Y

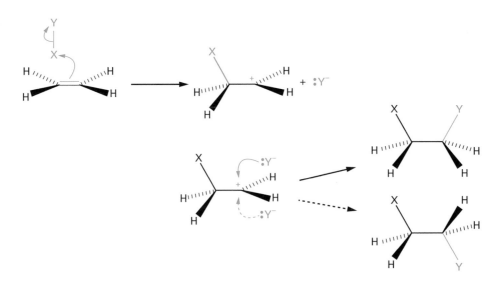

FIGURE 18.14 The electron pair in the pi bond forms a new bond between carbon and atom X. The bond between X and Y breaks, with Y now carrying a negative charge. The negative ion can then react with the carbocation.

- The nature of the intermediate in the addition of **bromine** is usually simplified pre-university to match the similar addition reaction with hydrobromic acid, so the intermediate is described as a carbocation. However, we now know that the intermediate is more accurately described as a bromonium ion, as we explore further in Taking the Next Step 18.4.

A Deeper Look 18.2 describes the bromine water test.

TAKING THE NEXT STEP 18.4

Why is a bromonium ion a better description of the intermediate?

Figure 18.15 shows how the positively charged carbon in a carbocation can be stabilized by interaction with an electron pair donated by the bromine atom (which has three lone pairs). The formation of the second bond provides additional stability. The resulting **bromonium ion** is a more accurate representation of the intermediate, as we can prove by direct observation (by NMR) of some bromonium ion intermediates.

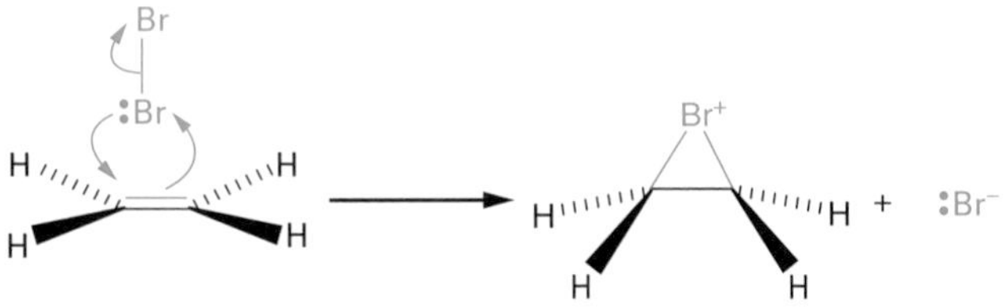

FIGURE 18.15 The formation of a bromonium ion relies on the formation of a second bond by electron-pair donation of one of bromine's three lone pairs.

Why is the structure of the bromonium ion important? It means that the bromide ion reacting in the second step **must** attack from below the three-membered ring, with consequences for the nature of the product of the reaction. For example, Figure 18.16 shows how (*E*)-but-2-ene gives an achiral product because the isomer formed by attack at either carbon atom has a plane of symmetry and hence is the *meso* isomer (Section 16.3).

By contrast, Figure 18.17 shows how an open carbocation would produce both the *meso* isomer and a chiral species. The fact that no chirality occurs provides further confirmation of the bromonium ion intermediate.

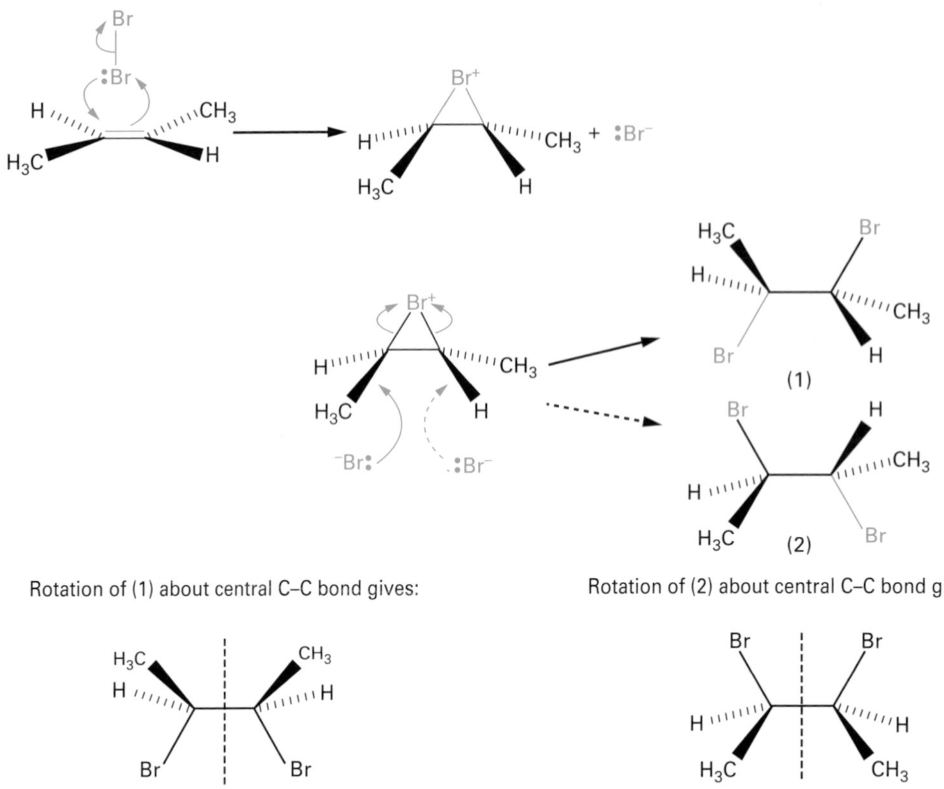

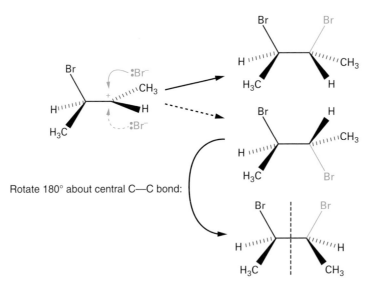

Rotation of (1) about central C–C bond gives:

Rotation of (2) about central C–C bond gives:

FIGURE 18.16 The bromonium ion formed from (*E*)-but-2-ene can only be attacked from below (at either of the two carbon atoms). In both cases, the molecule formed is the *meso* isomer (Section 16.3).

Rotate 180° about central C—C bond:

FIGURE 18.17 Attack from below on an open carbocation would give the *meso* isomer, but attack from *above* gives a chiral molecule.

18

A DEEPER LOOK 18.2

What happens with bromine water?

If **bromine *water*** is used, water can attack as a nucleophile in the second step in exactly the same way as water attacked under acidic conditions (Section 18.2). Reaction of bromine water with ethene produces 2-bromoethanol, $BrCH_2CH_2OH$, as shown in Figure 18.18. As 2-bromoethanol is colourless, the **alkene decolorizes 'orange' bromine water**.

This was an old-fashioned qualitative test for alkenes which works well but has been superseded by the use of spectroscopy, particularly NMR.

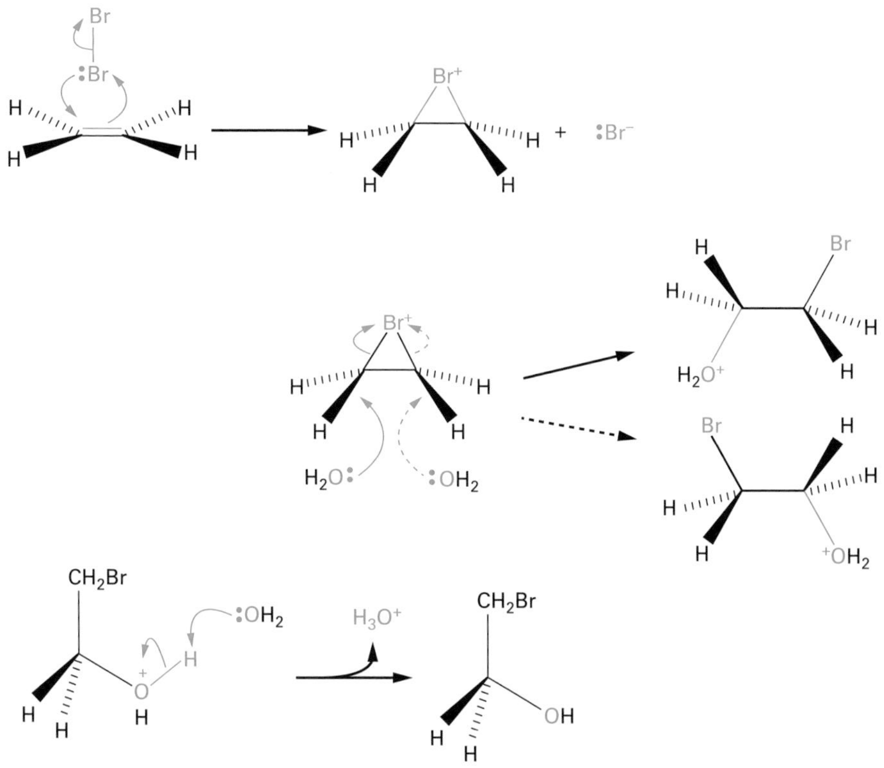

FIGURE 18.18 A water molecule can attack the bromonium ion. The molecule formed (after deprotonation) is 2-bromoethanol.

HYDROCARBONS: ARENES

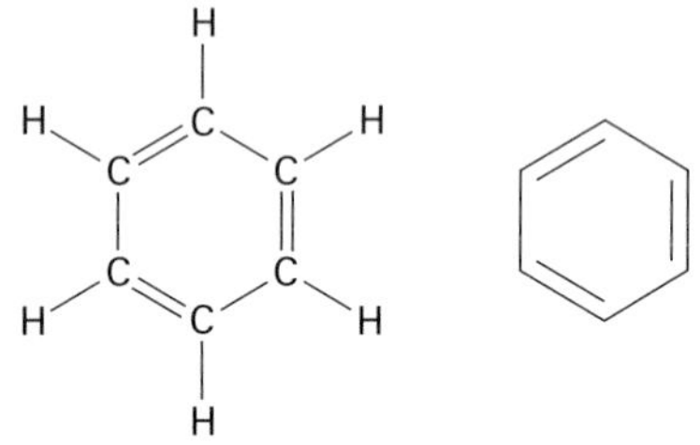

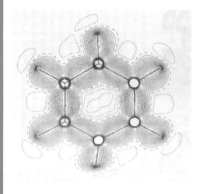

FIGURE 19.1 The Kekulé structure for benzene had alternating double and single bonds.

FIGURE 19.2 A diffraction image of benzene.

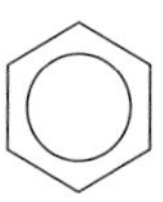

FIGURE 19.3 Another way of representing benzene, emphasizing the delocalization.

19.1 INTRODUCTION TO THE ARENES

- The archetypal **arene** (or **aromatic compound**) is **benzene**, whose formula is C_6H_6. The shape of the planar benzene molecule is a regular hexagonal ring of carbon atoms with one hydrogen atom bonded to each carbon atom.

- Figure 19.1 shows the original **Kekulé structure** for benzene, which had alternating double and single bonds.

- **Thermochemical evidence** from the hydrogenation enthalpy for benzene compared with the hypothetical molecule cyclohexa-1,3,5-triene suggests that benzene is more stable than expected from the Kekulé structure by about 150 kJ mol^{-1}. In addition, a number of experimental measurements prove that the bond lengths between neighbouring carbon atoms are all equal.

- Figure 19.2 shows a diffraction image of benzene, which proves that the **carbon–carbon bond lengths are all the same**: 139 pm (intermediate between a single bond and a double bond).

- The unexpected stability of benzene occurs because the π electrons (Section 2.1) are **delocalized** (Section 2.3) rather than localized between two carbon atoms. Figure 19.3 shows another way of representing benzene that emphasizes this delocalization. The **delocalization energy** is 150 kJ mol^{-1}. Taking the Next Step 19.1 explains this delocalization in more detail.

- Figure 19.4 shows that the resulting **electron density is symmetrical** round the ring and that there is **high electron density above and below the ring**, as indicated by the red coloration.

- The only reactions of benzene (apart from combustion) studied pre-university are **electrophilic substitution** reactions.

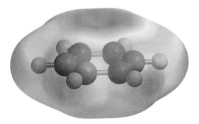

FIGURE 19.4 The electrostatic potential map for benzene: note the high electron density above (and below) the ring.

TAKING THE NEXT STEP 19.1

The delocalization in benzene in more detail

Figure 19.5 shows the **lowest-energy bonding molecular orbital**, which has all six atomic orbitals in phase. The two electrons occupying this orbital *bond all six carbon atoms together*. The equal bond lengths now seem entirely natural.

There are also four other electrons, which occupy **two other delocalized bonding molecular orbitals**. Figure 19.6 shows their shapes (which are rarely seen in pre-university textbooks). These other two bonding orbitals have one nodal plane (Section 2.1) each, which can lie either through two atoms or through two bonds, as shown in Figure 19.6. Computer calculations show that these two orbitals are degenerate (Section 1.4). The antibonding orbitals are not filled.

Some pre-university textbooks incorrectly claim that *all six* electrons are in the lowest-energy molecular orbital, which cannot happen as a maximum of two electrons can occupy any orbital according to the Pauli exclusion principle (Section 1.4).

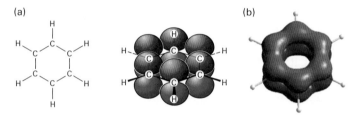

FIGURE 19.5 (a) Each carbon atom uses its 2s and two of its 2p atomic orbitals to form the **σ framework** of the benzene molecule. The remaining 2p atomic orbitals all combine together to form a set of delocalized π molecular orbitals. The six remaining valence electrons fill three bonding molecular orbitals, each of which is delocalized (together constituting a **delocalized π cloud**). (b) The lowest-energy orbital arises from overlap from all six orbitals in phase, Section 2.1. Thus the total *number* of bonds is correctly predicted by the Kekulé structure, but the detailed electron density is not.

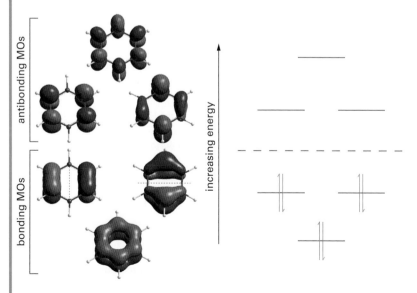

FIGURE 19.6 The six atomic p orbitals create six π molecular orbitals (MOs)— three bonding and three antibonding. The lowest-energy bonding MO has all p orbitals in phase. The dashed blue lines show the nodal planes in the other two bonding MOs. The highest-energy antibonding MO has all p orbitals out of phase.

19.2 ELECTROPHILIC SUBSTITUTION REACTIONS OF BENZENE

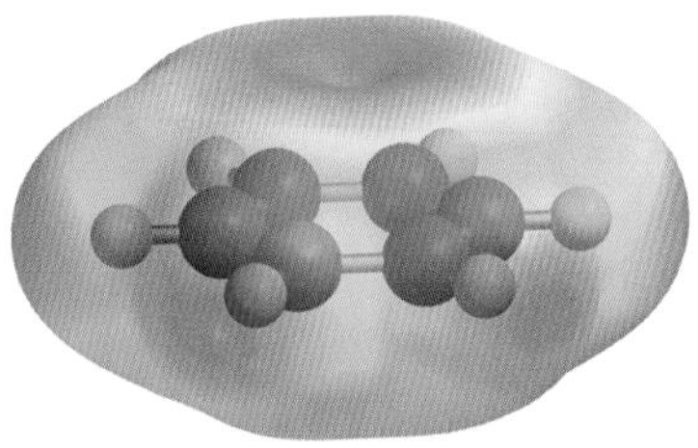

- Because of the high electron density above and below the benzene ring, as shown again in Figure 19.7, benzene reacts with electrophiles. This is similar to the situation with ethene (Figure 18.6).

- However, benzene's reaction with electrophiles, shown in Figure 19.8, differs from ethene's reaction in two important ways:
  - The reaction is slower due to the stable delocalized ring. The only electrophiles that react successfully are *positively charged* electrophiles E^+.
  - The intermediate ion that is formed after E^+ attaches *deprotonates* to reform the stable delocalized ring, so the overall process is **electrophilic substitution**.

- The structure of the intermediate needs careful thought. The carbon atom with the electrophile attached is tetrahedral as four atoms are bonded to it. It therefore has no other orbitals left and so is *not* involved in the delocalization, which now extends over only *five* carbon atoms. The **horseshoe** shape aims to illustrate this fact: the horseshoe should not be extended beyond the two carbon atoms nearest to the tetrahedral carbon atom. A Deeper Look 19.1 explains more about the intermediate.

FIGURE 19.7 The electrostatic potential map for benzene: note the high electron density above (and below) the ring.

We call the intermediate in electrophilic substitution the **Wheland intermediate**.

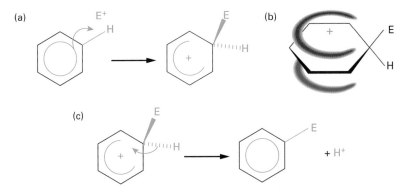

FIGURE 19.8 (a) The benzene π cloud attracts the E^+ ion: the ring provides an electron pair to form the new bond. (b) The remaining five carbon atoms share four electrons. (c) The intermediate loses a proton (deprotonates) to form the product, regaining the delocalization energy of the benzene ring.

A DEEPER LOOK 19.1

The Wheland intermediate

The Wheland intermediate is a cation that has four π electrons delocalized over five carbon atoms. A simpler cation called the allyl cation is frequently studied at university: the **allyl cation** has two π electrons delocalized over three carbon atoms. Figure 19.9 shows a cartoon of the three atomic orbitals (AOs) and the resulting computed molecular orbitals (MOs). Three AOs

19

produce three MOs (Taking the Next Step 2.1). Being the cation, there are only two electrons.

The lowest-energy (bonding) MO contains two electrons and bonds all three atoms together: one delocalized π bond. The middle orbital is a **non-bonding molecular orbital** (**NBMO**): there is no overlap between adjacent orbitals. The NBMO is the orbital from which the electron is removed in making the cation. *Therefore the positive charge on the cation is shared equally by the outer atoms with* zero *charge on the central atom.* The highest-energy (antibonding) orbital is unoccupied.

A simple open-chain model for the Wheland intermediate is similar but with two more carbon atoms together with their orbitals and electrons. Five AOs produce five MOs. Figure 19.10 shows a cartoon of the two bonding

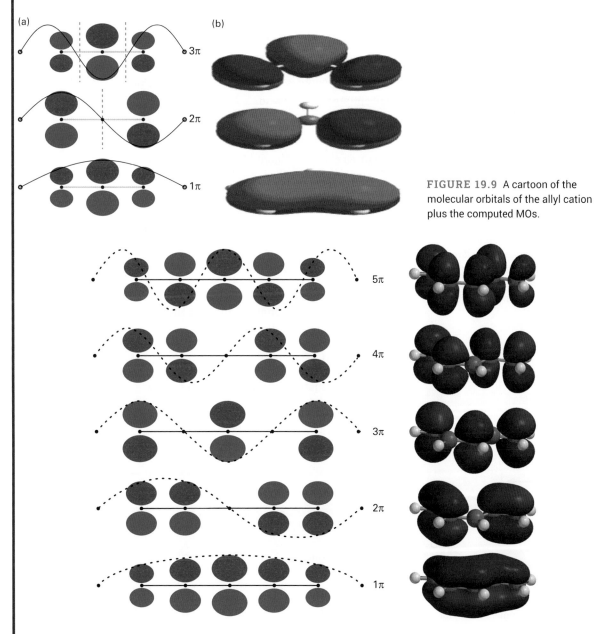

(a)

(b)

3π

2π

1π

FIGURE 19.9 A cartoon of the molecular orbitals of the allyl cation plus the computed MOs.

5π

4π

3π

2π

1π

FIGURE 19.10 A cartoon of the molecular orbitals of the open-chain model for the Wheland intermediate plus the computed MOs.

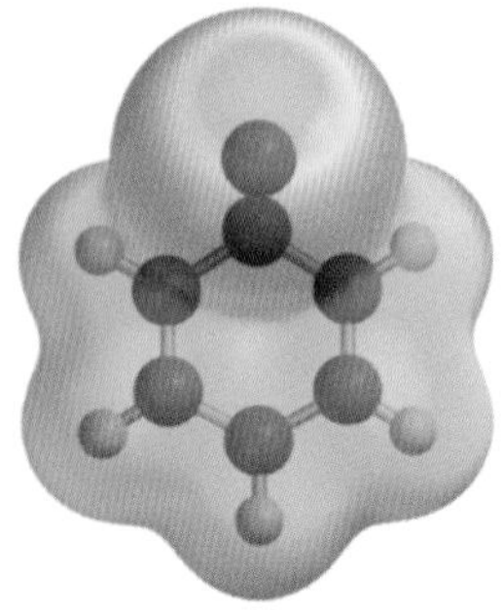

FIGURE 19.11 The electron density in the Wheland intermediate is not evenly distributed. The blue colour in this electrostatic potential map is much more intense at the 2-, 4-, and 6- positions compared with the 3- and 5- positions.

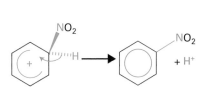

FIGURE 19.12 The nitronium ion is linear, just like the isoelectronic CO_2. The dark blue coloration in the electrostatic potential map shows that the positive charge is largely on the nitrogen atom. (Remember that the Spartan software does not show multiple bonding, Figure 18.6.)

19

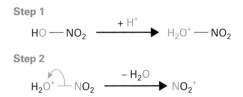

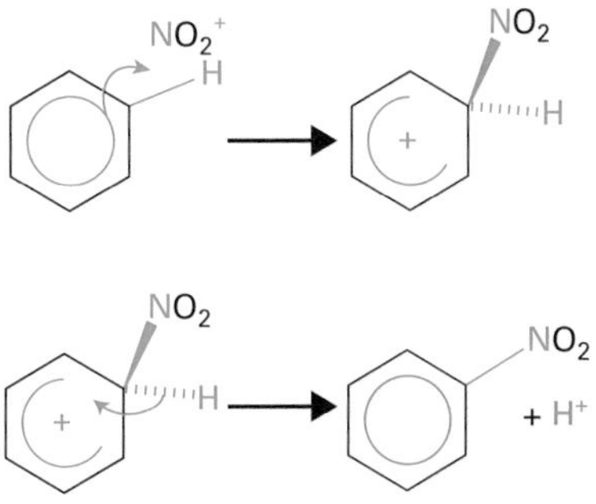

FIGURE 19.14 The mechanism for the nitration of benzene.

MOs and two antibonding MOs, together with the NBMO. The two bonding orbitals are each occupied by two electrons, forming two delocalized π bonds (the antibonding orbitals are again unoccupied). Again, the electron is removed from the NBMO.

The positive charge on the Wheland intermediate is shared equally by the outermost atoms and the central atom, with zero charge on the other two atoms. This is confirmed by looking at the different intensities of the blue colour in the electrostatic potential map of the Wheland intermediate, as seen in Figure 19.11.

19.3 ELECTROPHILIC SUBSTITUTION: NITRATION

• A particularly important reaction undergone by benzene, both in the laboratory and in industry, is **nitration**.

• A **nitrating mixture** is a mixture of concentrated nitric acid and concentrated sulfuric acid.

    – The nitrating mixture forms the electrophile that then reacts with benzene: the **nitronium ion NO_2^+** (which is isoelectronic, Section 2.4, with CO_2), as shown in Figure 19.12. Figure 19.13 describes its formation.

• The overall reaction is

$$HNO_3 + 2H_2SO_4 \rightarrow NO_2^+ + H_3O^+ + 2HSO_4^-$$

Pre-university you may have been allowed to write a simpler equation with H_2O as one product, which would be protonated in the presence of the strong acid sulfuric acid.

The nitronium ion can be thought of as forming in two steps, Figure 19.13.
Step 1 Sulfuric acid protonates nitric acid HNO_3 (more correctly written as $HONO_2$, see A Deeper Look 13.1):

$$HONO_2 + H_2SO_4 \rightarrow H_2O^+ - NO_2 + HSO_4^-$$

Step 2 The protonated nitric acid then loses water:

$$H_2O^+ - NO_2 \rightarrow NO_2^+ + H_2O$$

Step 1
$$HO - NO_2 \xrightarrow{\ + H^+\ } H_2O^+ - NO_2$$

Step 2
$$H_2O^+ - NO_2 \xrightarrow{\ - H_2O\ } NO_2^+$$

FIGURE 19.13 The formation of the nitronium ion.

The NO_2^+ is a powerful electrophile because it has a full positive charge.

    Figure 19.14 shows the mechanism of nitration, which follows the generic one shown in Figure 19.8.

- The product of the reaction is $C_6H_5NO_2$, **nitrobenzene**.

- Nitration (under harsher industrial conditions) is used to produce **explosives** such as trinitromethylbenzene (trinitrotoluene, TNT). After reduction of nitrobenzene (using Sn/HCl, Section 24.3) to phenylamine (aniline), we can make **dyes** and **drugs**.

19.4 ELECTROPHILIC SUBSTITUTION: FRIEDEL–CRAFTS ACYLATION

- The reagents for **Friedel–Crafts acylation** are an acyl chloride (Section 23.3), **RCOCl**, and aluminium chloride **AlCl₃**, which acts as a **Lewis acid** (Section 7.6).

  - The Lewis acid catalyst generates the electrophile RCO⁺, which we call an **acylium ion**.

  - Figure 19.15 shows an acylium ion: the figure also explains why the carbonyl carbon is the atom that becomes bonded to the benzene ring.

$$RCOCl + AlCl_3 \rightarrow RCO^+ + AlCl_4^-$$

- Figure 19.16 shows the mechanism of Friedel–Crafts acylation, which follows the generic one shown in Figure 19.8.

  - The product of the reaction is C_6H_5COR. If R is a methyl group as in Figure 19.16, the product is $C_6H_5COCH_3$, **phenylethanone** (acetophenone).

  - Friedel–Crafts acylation is an important reaction in industry; for example, it is the first step in one synthesis of the pain-killer ibuprofen.

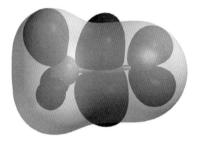

FIGURE 19.15 The grey semitransparent layer shows the shape of the electron density of the acylium ion CH_3CO^+. Within this is the structure of the lowest unoccupied molecular orbital (LUMO, A Deeper Look 16.3). In the LUMO the highest electron density (which is proportional to the square of the size of the lobes) occurs on the carbonyl carbon, which therefore becomes bonded to the benzene ring. (This software package shows the double bond between the carbon and oxygen atoms.)

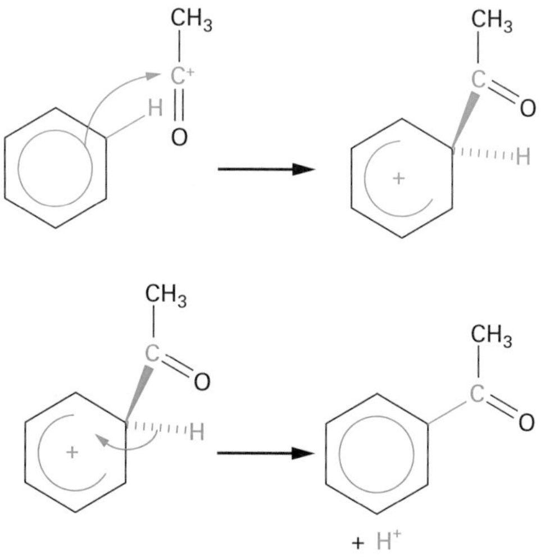

FIGURE 19.16 The mechanism for the Friedel–Crafts acylation of benzene.

19.5 ELECTROPHILIC SUBSTITUTION: HALOGENATION

Metallic iron can also serve as a catalyst because it reacts with chlorine to form FeCl$_3$.

- Benzene requires a catalyst for halogenation (reaction with a halogen such as Cl$_2$) to be successful. **The catalyst is again a Lewis acid** (Section 7.6) such as FeCl$_3$ (we can also use AlCl$_3$). The catalyst generates the Cl$^+$ electrophile:

  $-$ Cl$_2$ + FeCl$_3$ → Cl$^+$ + FeCl$_4^-$

  > The presence of an actual positive ion is not certain; it may be that the electrophile is a heavily-partially-positively charged chlorine, still coordinated to the iron.

- Figure 19.17 shows the mechanism of chlorination, which follows the generic one shown in Figure 19.8.

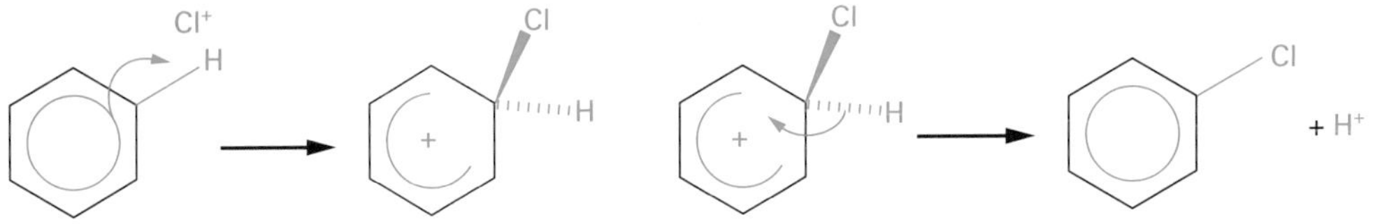

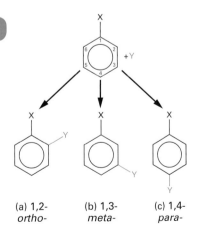

FIGURE 19.17 The mechanism for the chlorination of benzene.

The product of the reaction is C$_6$H$_5$Cl, **chlorobenzene**.

19.6 DIRECTIVE EFFECTS

Pre-university you may have learned about the effect on the position at which a second electrophile attaches because of the presence of an existing group on the benzene ring. Taking the Next Step 19.2 explains this **directive effect** of an existing group (such as an alkyl group).

FIGURE 19.18 The three possible isomers of a disubstituted benzene compound.

(a) 1,2-
ortho-

(b) 1,3-
meta-

(c) 1,4-
para-

TAKING THE NEXT STEP 19.2

An alkyl group speeds up reaction and directs ortho *and* para

If there is an alkyl group already on the benzene ring, it will play a very important role in the distribution of the possible isomers that can be formed. Figure 19.18 explains the notation *ortho-*, *meta-*, and *para-* for disubstituted benzene rings.

The **electron-donating** (Section 18.3) methyl group stabilizes the Wheland intermediate, so methylbenzene (toluene) reacts more quickly than benzene itself. Another consequence is a very significant increase in the amount of the *ortho-* (*o-*) and *para-* (*p-*) isomers relative to the *meta-* (*m-*) isomer because the methyl group preferentially stabilizes the Wheland intermediate that produces these two isomers, as shown in Figure 19.19. (An **electron-withdrawing** group, such as the nitro group, makes the ring react more slowly and directs the electrophile preferentially to the *meta-* position.)

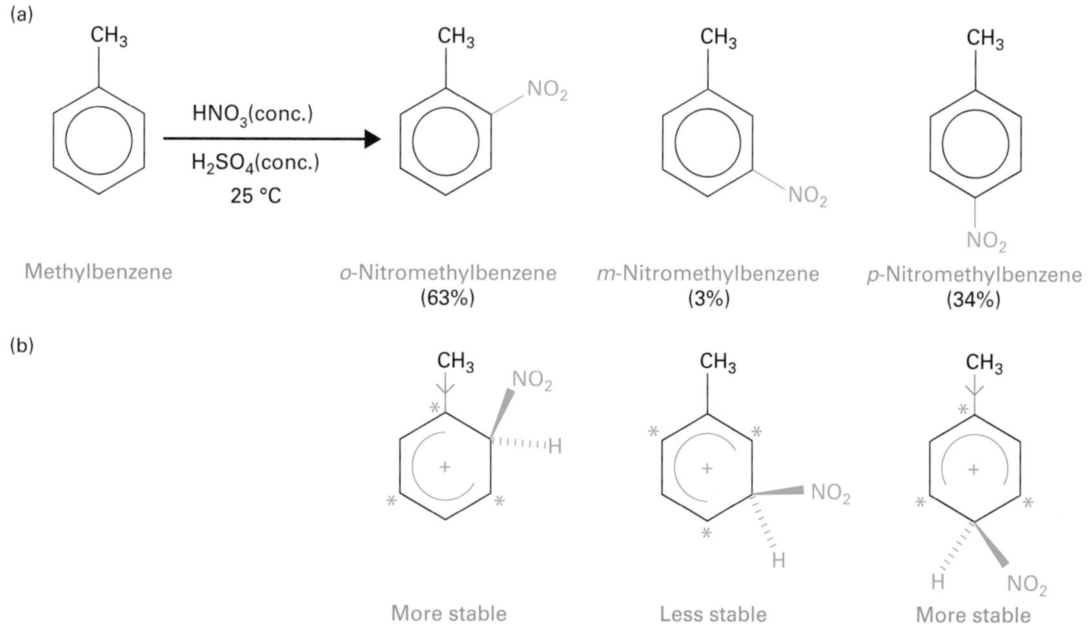

FIGURE 19.19 (a) The nitration of methylbenzene produces more of the *o*- and *p*- isomers than of the *m*- isomer. (b) The Wheland intermediates leading to *o*-nitromethylbenzene, *m*-nitromethylbenzene, and *p*-nitromethylbenzene. The positions at which the positive change is concentrated are labelled with a star (positions *ortho* and *para* relative to the incoming –NO$_2$ group). The –CH$_3$ group donates electron density (shown by the green arrow) to the ring and therefore preferentially stabilizes the two intermediates with an adjacent positive charge.

A Deeper Look 19.1 explains the reason why the positive charge in the Wheland intermediate is only shared by three of the five carbons.

The *ortho/para* ratio varies considerably depending on the alkyl group, so the above figures refer specifically to the nitration of methylbenzene (toluene). If the alkyl group is *t*-butyl, the nitration product is 82% *para* due to the much larger steric effect of the *t*-butyl group.

Taking the Next Step 19.3 explains the consequence of this for the **Friedel–Crafts alkylation** reaction. A Deeper Look 19.2 provides an alternative explanation for *ortho/para* direction. A Deeper Look 19.3 explains one other stabilization mechanism that can account for *ortho/para* direction.

TAKING THE NEXT STEP 19.3

What is Friedel–Crafts alkylation?

There is a **Friedel–Crafts alkylation** reaction, which is very similar to the Friedel–Crafts acylation reaction (Section 19.4). It uses a chloroalkane, RCl, in place of the acyl chloride, RCOCl:

$$RCl + AlCl_3 \rightarrow R^+ + AlCl_4^-$$

However, the reaction is *much* less easy to control because once we have added one alkyl group, the alkylbenzene reacts faster than benzene does, so *polysubstitution* is common. (Taking the Next Step 19.2 explains the effect of an alkyl group already present on the ring.)

The original carbocation R$^+$ can also rearrange. For example, CH$_3$CH$_2$CH$_2^+$ can rearrange to the more stable carbocation (Section 18.3) (CH$_3$)$_2$CH$^+$.

It is also possible to use the corresponding bromine compound RBr and a Lewis acid such as AlBr$_3$.

19

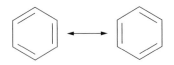

FIGURE 19.20 The double-headed arrow indicates the averaging of the two **resonance structures** (which are also called **canonical structures**). It is **very** important not to confuse this *one double*-headed arrow with the *two half*-arrows used for an equilibrium reaction (Section 6.1).

A DEEPER LOOK 19.2

How is *ortho/para* direction explained using 'resonance'?

We have already explained (Section 19.1) that the original Kekulé structure for benzene cannot explain benzene's symmetrical structure. Before the concept of delocalization was understood, one solution proposed was that an average of two alternative Kekulé structures would be better, as shown in Figure 19.20. A double-headed arrow is used to show this averaging: the term used to describe this averaging is '**resonance**'. (Remember that neither alternative resonance structure is adequate on its own.) This use of the term bears no relation to the meaning of the word in nuclear magnetic resonance (NMR, Section 26.3) or indeed in physics. However, you will frequently come across this use of the term 'resonance' in organic textbooks.

Each alternative resonance structure seeks to identify **two**-centre two-electron bonds, thus ignoring the fact that **six**-centre two-electron delocalized bonds exist (Figure 19.5). Resonance does **not** imply an oscillation between the two structures: rather, the 'real' structure for benzene is the average of the two resonance structures. For example, the average number of bonds between each pair of carbon atoms is 1.5 bonds.

The resonance description can be used to explain the *ortho/para* direction in methylbenzene (toluene), which is discussed in Taking the Next Step 19.2. To do so, three alternative resonance structures are needed, as shown in Figure 19.21 for the *ortho* isomer.

None of the three resonance structures is adequate on its own. The resonance structure with the positive charge on the carbon bonded to the methyl group is the most stable, as this is a tertiary carbocation (Section 18.3).

A similar description can be given for the *para* isomer, as shown in Figure 19.22. Once again, the resonance structure with the positive charge on the carbon bonded to the methyl group is the most stable, as this is a tertiary carbocation.

The Wheland intermediate leading to the *meta* isomer has no resonance structure that involves a tertiary carbocation, and so is significantly less stable.

Notice how the average of the three resonance structures in Figures 19.21 and 19.22 places the positive charge on the same atoms as in the delocalization model (Figure 19.19).

Molecular orbital (MO) theory provides a much more satisfactory quantitative description. The resulting MOs are all delocalized; the lowest-energy bonding MO in benzene itself is delocalized over all six carbon atoms (Figure 19.5).

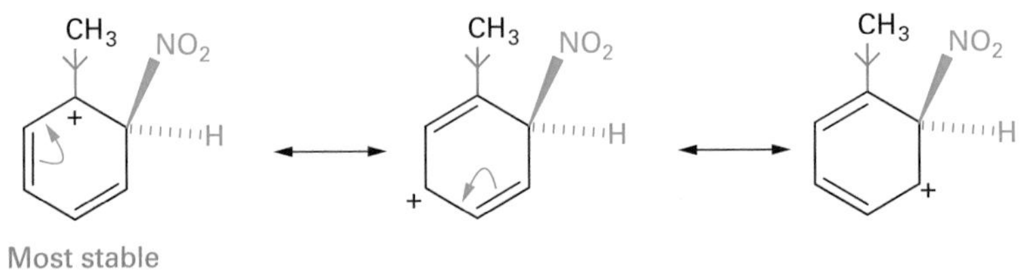

Most stable

FIGURE 19.21 The Wheland intermediate leading to the *ortho* isomer. The position of the positive charge can be shifted from one resonance structure to another by the motion of an electron pair.

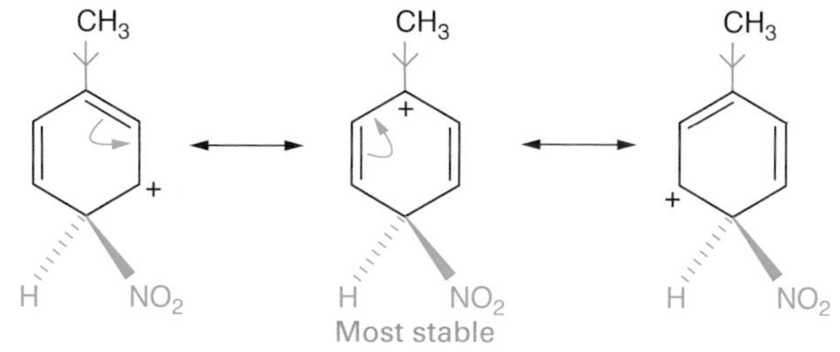

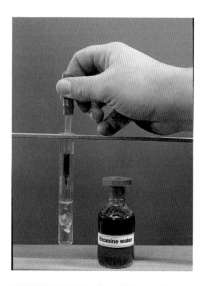

FIGURE 19.22 The Wheland intermediate leading to the *para* isomer. The position of the positive charge can be shifted from one resonance structure to another by the motion of an electron pair.

A DEEPER LOOK 19.3

Ortho/para direction in phenol and phenylamine

An alternative way in which the intermediates leading to the *ortho-* and *para-* isomers can be stabilized is by delocalization of the charge in the Wheland intermediate out onto the group already present on the ring. (This can happen if the group has a lone pair to contribute to the delocalized structure.) This stabilization also produces a *much* more rapid reaction than for any of the other substituted benzenes. This stabilization occurs with an –OH group (in the molecule phenol) or an –NH$_2$ group (in the molecule phenylamine/aniline). Figure 19.23 illustrates the bromination of phenol, which very rapidly produces the molecule 2,4,6-tribromophenol, shown in Figure 19.24.

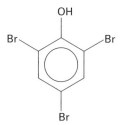

FIGURE 19.24 2,4,6-tribromophenol. The equivalent chlorine molecule, 2,4,6-trichlorophenol, is called TCP. TCP is a well-known antiseptic.

FIGURE 19.23 The –OH group in phenol C$_6$H$_5$OH activates the ring to substitution. Phenol decolorizes bromine water (remember that benzene does not); the product is an immediate white precipitate of 2,4,6-tribromophenol.

19

HALOGENOALKANES

20.1 INTRODUCTION TO THE HALOGENOALKANES

- Figure 20.1 shows that **halogenoalkanes** (also called **haloalkanes**; you will frequently come across the name **alkyl halides** at university) have carbon–halogen bonds that are significantly polar (Section 2.2): as a result, the carbon atom is partially positively charged. This makes the **carbon atom susceptible to nucleophilic attack** (Section 16.4).

- One important characteristic type of reaction of halogenoalkanes is therefore **nucleophilic substitution**.

  - The carbon–halogen bond breaks and the halide ion (which is a good **leaving group**) is released.

    The carbon–fluorine bond is very strong as a result of the 2p–2p orbital overlap: it is even stronger than the carbon–oxygen bond. (For the other halogens the orbital overlap is not as significant since the halogen electrons are in a higher orbital than the carbon electrons.) This means that the fluoride ion is a bad leaving group and fluoroalkanes react very poorly.

  - The rate-limiting step (Section 10.3) involves bond-breaking (either completely or partially). Therefore the **carbon–halogen bond enthalpy**, shown in Table 20.1, determines the relative rate of reaction: chloroalkanes react more slowly than bromoalkanes and iodoalkanes are fastest. The carbon-iodine bond is the longest and hence weakest bond.

(a)   (b)   (c)

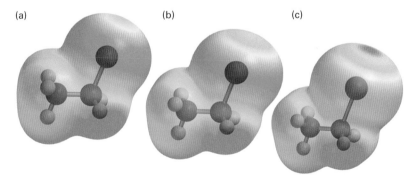

FIGURE 20.1 Electrostatic potential maps of the halogenoethanes. Note that the atomic radii are: Cl, 0.099 nm; Br, 0.114 nm; I, 0.133 nm. (a) The C–Cl bond (length 0.177 nm). (b) The C–Br bond (length 0.193 nm). (c) The C–I bond (length 0.214 nm). The red colour indicates the partial negative charge (Section 2.2) on the halogen (itself the coloured atom in each molecule), which decreases from Cl to I.

TABLE 20.1 Average (mean) bond enthalpies

Bond	Average bond enthalpy/kJ mol^{-1}
C–F	484
C–Cl	338
C–Br	276
C–I	238

Note that because iodine has the lowest electronegativity of all the halogens the polarity of the carbon–iodine bond is the lowest, as seen in Figure 20.1. If the polarity of the bond was the more important factor affecting reactivity, iodoalkanes would react slowest. Hence the bond enthalpy is the more important factor for explaining the higher rate of the reaction. Note, however, that other factors do matter (such as the solvent).

20.2 NUCLEOPHILIC SUBSTITUTION REACTIONS OF HALOGENOALKANES

- Three common **nucleophilic substitution** reactions of halogenoalkanes (RX, R being an alkyl group and X a halogen) are those with sodium hydroxide, potassium cyanide, and ammonia.

- **Sodium hydroxide** (hydroxide ion being the nucleophile) reacts in **aqueous solution** and causes hydrolysis of the halogenoalkane, forming an **alcohol**, ROH:
  - RX + NaOH → ROH + NaX

- **Potassium cyanide** (cyanide ion being the nucleophile) is typically used in a solvent mixture consisting of **aqueous ethanol**. Reaction produces a **nitrile**, RCN:
  - RX + KCN → RCN + KX

- **Excess ammonia reacts in a sealed tube** to produce an **amine**, RNH_2:
  - RX + $2NH_3$ → RNH_2 + NH_4X
    - However, the resulting amine is also nucleophilic and hence other products can form by subsequent reaction (Section 24.3): R_2NH, R_3N, and R_4N^+ X^-.

- The most likely mechanism for nucleophilic substitution depends on a number of factors, especially the detailed carbon skeleton, as shown in Figure 20.2. We classify halogenoalkanes as follows:
  - A **primary halogenoalkane** has the structure RCH_2X: it has one alkyl group attached to the carbon bonded to the halogen.
  - A **secondary halogenoalkane** has the structure RR'CHX: it has two alkyl groups attached to the carbon bonded to the halogen.

(a)

Bromoethane, a primary halogenoalkane

(b)

2-Bromopropane, a secondary halogenoalkane

(c)

2-Bromo-2-methylpropane, a tertiary halogenoalkane

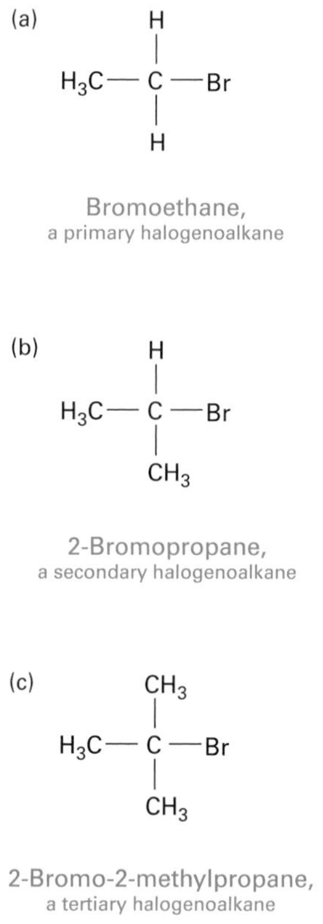

FIGURE 20.2 The classes of halogenoalkanes.

(a)

(b)

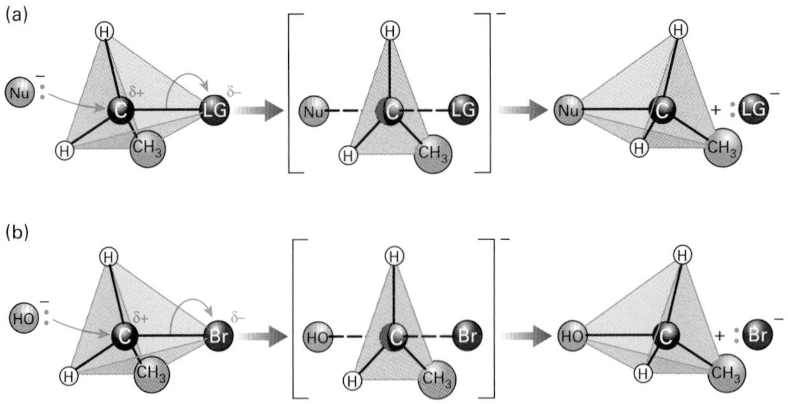

FIGURE 20.3 (a) A nucleophile Nu:⁻ is attracted to the δ+ carbon atom of the C–X bond. It approaches from the opposite side to the halogen, where its attack is not impeded by the bulky halogen with its δ– charge. The non-organic product of the reaction is the leaving group :LG⁻. In the transition state, both Nu and LG are *partially* bonded to carbon. The transition state involves a p orbital on the carbon atom, and so forces a 'reverse-side' attack by the incoming nucleophile. (b) shows the specific example of hydroxide ion as the nucleophile.

At pre-university level, it is often possible to score full marks without giving any indication of the direction of attack. This can lead to a lack of understanding of the geometrical consequences; see the in-text comment for example.

At university, this is called the **S$_N$2 mechanism**, which stands for substitution, nucleophilic, bimolecular (because the transition state involves two species).

– A **tertiary halogenoalkane** has the structure RR'R"CX: it has three alkyl groups attached to the carbon bonded to the halogen.

• A **primary halogenoalkane** is most likely to react by the mechanism shown in Figure 20.3. This is the mechanism you will probably have studied pre-university, usually with hydroxide ion as the nucleophile.

> If the starting material has an asymmetric carbon atom (Section 16.3), an inversion of the structure around the carbon atom occurs, which we call **Walden inversion**. Reaction of the chiral molecule (+)-2-iodooctane with radioactive iodide ion gave a rate of racemization exactly twice the rate of incorporation of radioactivity, as each and every incorporation of a radioactive atom inverted a (+) isomer into a (−) isomer.

There is plenty of kinetic data to support this mechanism. For example, in the case of bromoethane CH_3CH_2Br, the rate equation is

$$rate = k[CH_3CH_2Br][OH^-].$$

Notice how the rate depends on the concentration of both the halogeno-alkane *and* the hydroxide ion. Figure 20.4 shows the intermediate stages in the similar reaction with cyanide ion as the nucleophile. These calculations confirm the 'reverse-side' attack direction.

• A **tertiary halogenoalkane** is most likely to react by the **S$_N$1 mechanism**, which stands for substitution, nucleophilic, unimolecular (because the transition state involves only one species). Taking the Next Step 20.1 discusses the S$_N$1 mechanism, which you are less likely to have studied pre-university.

20

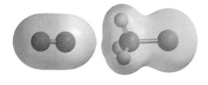

The carbon atom of the cyanide ion approaches the halogenomethane.

FIGURE 20.4 Five stages in the S$_N$2 reaction between cyanide ion and a halogenomethane.

As cyanide ion gets closer, orbital overlap starts to occur.

In the transition state, both carbon and the halogen are partially bonded.

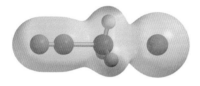

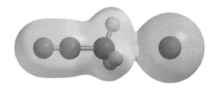

The carbon-halogen bond is almost completely broken

Notice the movement of the three hydrogens on the carbon from facing left to vertical to facing right.

TAKING THE NEXT STEP 20.1

What is the S$_N$1 mechanism?

Comparing the structure of a primary and a tertiary halogenoalkane shows that the reaction mechanisms are likely to be different. The δ+ carbon atom in a tertiary halogenoalkane is surrounded by large alkyl groups, which obstruct the attack of the nucleophile. We call this effect **steric hindrance**. The hydrolysis of a tertiary halogenoalkane proceeds as in Figure 20.5.

There is kinetic evidence for this mechanism in the case of 2-bromo-2-methylpropane (*tertiary*-butyl bromide, *t*-BuBr, Section 16.2). The rate equation is simply *rate* = k[*t*-BuBr]. Notice how the rate-limiting step only involves the halogenoalkane. Measuring the rate of reaction as a function of the sodium hydroxide concentration shows that, whatever the concentration of NaOH, the rate is the same. The S$_N$1 mechanism occurs in two steps, whereas the S$_N$2 mechanism is all happening at the same time, which we describe as 'concerted'.

20

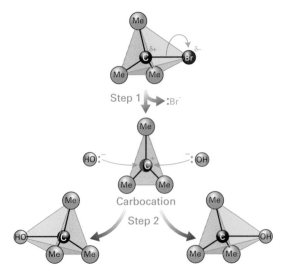

Step 1. $\Rightarrow$:Br⁻

Carbocation
Step 2

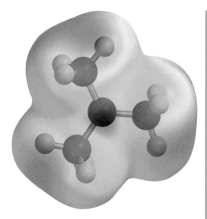

FIGURE 20.6 The electrostatic potential map for the carbocation $(CH_3)_3C^+$.

A **secondary** halogenoalkane can react by either S_N2 or S_N1. A Deeper Look 20.1 discusses the factors that affect the likelihood of each mechanism occurring.

FIGURE 20.5 Step 1. In the course of random collisions, typically with solvent molecules, the halogenoalkane ionizes by heterolytic fission. The result of this rate-limiting step is a carbocation and a halide ion:

$$(CH_3)_3C-Br \rightarrow (CH_3)_3C^+ + Br^-$$

Step 2. The carbocation attracts a hydroxide ion OH^- and bonds to it, forming the product. This step happens much faster than step 1, so the overall rate does not depend on the hydroxide ion concentration. Notice how the S_N1 mechanism can either invert or retain the structure around an asymmetric carbon atom (Section 16.3), as the nucleophile can attack from either side of the *planar* carbocation, see Figure 20.6.

A DEEPER LOOK 20.1

S_N2 or S_N1?

The factors that affect the relative rates of these two mechanisms are the carbon skeleton, the nucleophile, the solvent, the leaving group (and the temperature).

In most situations, the most important factor is the structure of the **carbon skeleton**, as already explained. Additional alkyl groups cause greater steric hindrance (Taking the Next Step 20.1), which slows the S_N2 mechanism, *and*, being electron-donating, help to stabilize the carbocation in S_N1, making that mechanism faster; see Table 20.2.

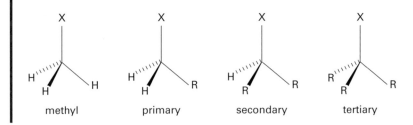

methyl   primary   secondary   tertiary

TABLE 20.2 The usual nucleophilic substitution mechanism shown by the different classes of halogenoalkanes.

Class	Methyl	Primary	Secondary	Tertiary
S_N1 reaction?	very unlikely	unlikely	moderate	excellent
S_N2 reaction?	very good	good	moderate	unlikely

The concentration of the **nucleophile** has no effect on the relative rate of S_N1 reactions (Taking the Next Step 20.1). However, increasing concentration of the nucleophile increases the rate of S_N2 reactions.

For S_N2 reactions, negatively charged nucleophiles are usually more reactive than neutral ones. For the same attacking atom, nucleophilicity roughly parallels basicity.

Thirdly, the important interaction is between the lone pair of the nucleophile and the C–X σ* antibonding orbital. Antibonding orbitals are higher in energy than lone pairs, so the higher the energy of the lone pair, the better the overlap. Therefore, nucleophiles from lower down in the periodic table, which have higher-energy lone pairs, give faster S_N2 reactions. Compare I⁻ with Br⁻ or PhS⁻ with PhO⁻, as shown in Table 20.3.

The C–X σ* antibonding orbital is the LUMO (A Deeper Look 16.3) of the halogenoalkane.

The **solvent** can play a major role in assisting S_N1 reactions because the transition state is necessarily more polar than the starting halogenoalkane. Therefore, a more polar solvent will increase the chances of this mechanism occurring by solvating the ions formed (remember that the hydration enthalpies of ions are exothermic, Section 5.4). Water and/or methanol is an ideal solvent. The role of the solvent in S_N2 is more complicated.

A good **leaving group** speeds up both S_N2 and S_N1 reactions so makes little difference to which dominates. Good leaving groups include stable anions such as bromide and iodide ions. As mentioned previously, fluoride ion is a bad leaving group. We can understand this by noting that HF is a weak acid (Section 7.3), with little ionization to form H⁺ and F⁻. A neutral leaving group is much better than its conjugate base (Section 7.1). This is particularly important for the reaction of alcohols, which do not react with halide ions (Section 20.2) because hydroxide ion is a very bad leaving group. However, reaction occurs well in strong acid, which protonates the alcohol to form ROH_2^+, creating the much better leaving group H_2O.

S_N2	S_N1
Concerted bimolecular mechanism	Two-step unimolecular mechanism
Kinetics second-order	Kinetics first-order
Asymmetric carbon atoms invert	Asymmetric carbon atoms racemize
Primary fastest, tertiary slowest	Tertiary fastest, primary slowest

TABLE 20.3 The rates of reaction of various nucleophiles with bromomethane in ethanol (relative to water).

nucleophile	F⁻	H_2O	Cl⁻	Et_3N	Br⁻	PhO⁻	EtO⁻	I⁻	PhS⁻
relative rate	0.0	1.0	1100	1400	5000	2.0×10^3	6×10^4	1.2×10^5	5.0×10^7

20.3 ELIMINATION REACTIONS OF HALOGENOALKANES

If the halogenoalkane already has a double bond involving the carbon attached to the halogen, an alkyne (Section 18.1) will be formed. **Alkynes** have a carbon–carbon triple bond (Section 2.1).

• A second important characteristic type of reaction of halogenoalkanes is **elimination**, which produces an alkene.

  – For example, 2-bromopropane $(CH_3)_2CHBr$ and KOH react to form the alkene propene $CH_3CH=CH_2$ (plus KBr and H_2O).

• Pre-university, we can achieve elimination most easily by reaction with **hot concentrated alcoholic potassium hydroxide**.

> At university other bases are also frequently used, such as sodium ethoxide $NaOCH_2CH_3$ in ethanol.

• In elimination, the hydroxide ion is acting as a *base* (Section 7.1) rather than as a *nucleophile*. **Elimination will therefore normally compete with nucleophilic substitution.**

• Figure 20.7 shows the most common elimination mechanism studied pre-university. A Deeper Look 20.2 discusses the orbitals involved.

At university we call this the **E2** (elimination, bimolecular) **mechanism**. It is bimolecular because both the base and the organic compound are involved in the rate-limiting step.

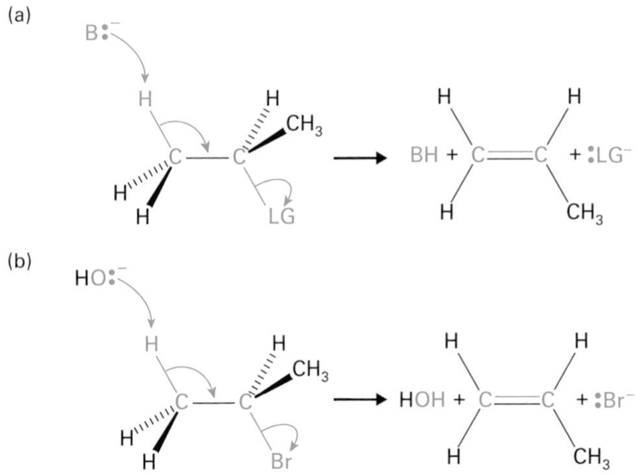

FIGURE 20.7 (a) The base B:⁻ removes a proton, the electron pair completes the double bond, and the leaving group :LG⁻ is released. (b) The specific example of hydroxide ion as the base.

Figure 20.7 (b) shows that elimination of 2-bromopropane using potassium hydroxide in ethanol produces propene.

A DEEPER LOOK 20.2

What are the orbitals involved in E2?

Orbital overlap is usually maximized in E2 elimination if the hydrogen and the halogen are aligned in one particular way, as shown in Figure 20.8. We describe this particular orientation in which they are opposite each other and lie in the same plane as **anti-periplanar**. In Figure 20.8, the view shown in the boxes (called a **Newman projection**) looks down the carbon–carbon bond direction.

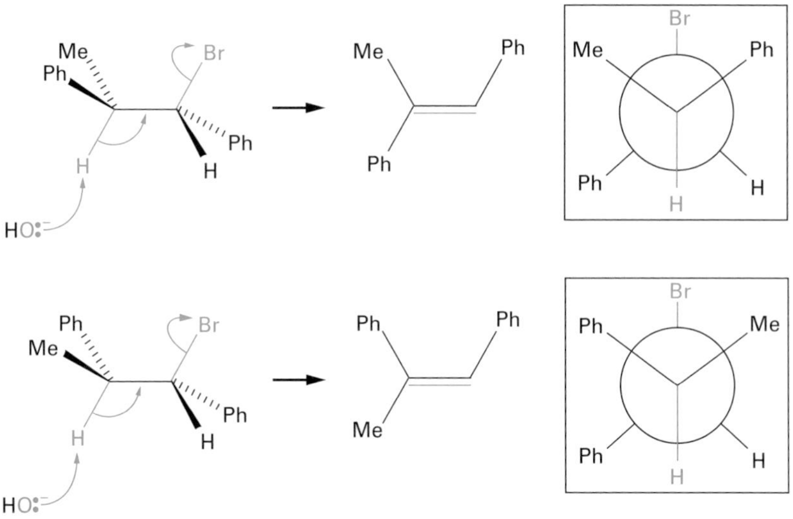

FIGURE 20.8 We call the best orientation for E2 elimination anti-periplanar. We call the view shown in the boxes a Newman projection.

- The elimination mechanism shown in Figure 20.8 involves the hydroxide ion in the rate-limiting step, so the concentration of the hydroxide ion will be an important factor in deciding the likelihood of elimination.

- Some halogenoalkanes can eliminate to produce **two products**.

  - Figure 20.9 shows that 2-bromobutane can eliminate to form but-1-ene if the hydrogen from carbon atom 1 is removed.

  - Figure 20.10 shows that 2-bromobutane can eliminate to form but-2-ene if the hydrogen from carbon atom 3 is removed. (*E* and *Z* isomers of but-2-ene can be formed.)

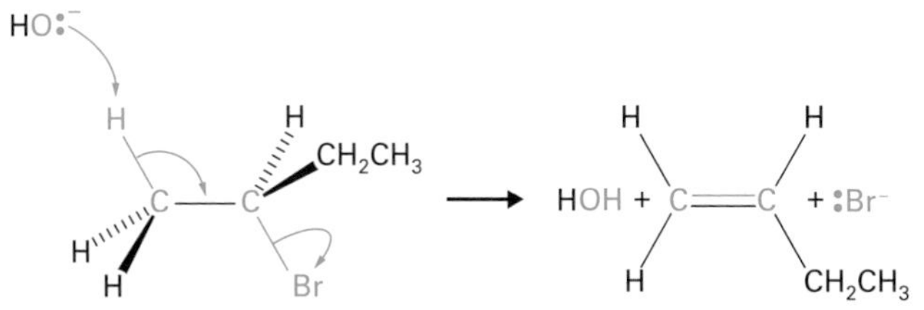

FIGURE 20.9 Removal of the proton from carbon atom 1 of 2-bromobutane forms but-1-ene.

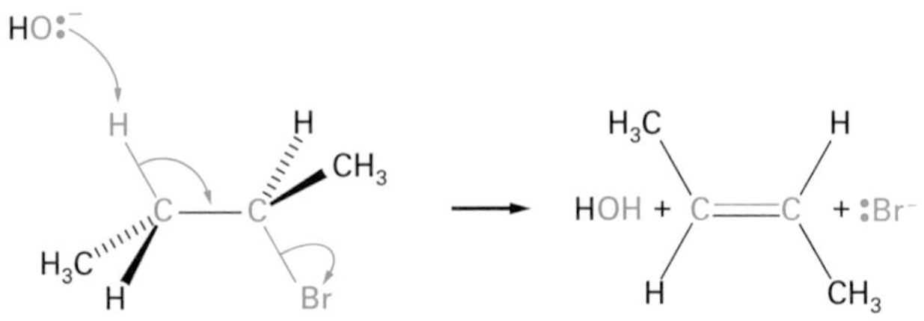

FIGURE 20.10 Removal of the proton from carbon atom 3 of 2-bromobutane forms but-2-ene.

20

Taking the Next Step 20.2 explains why tertiary halogenoalkanes usually eliminate via the alternative E1 mechanism. Taking the Next Step 20.3 discusses the factors that favour elimination over substitution.

TAKING THE NEXT STEP 20.2

What is the E1 mechanism?

The rate-limiting step in an **E1 (elimination, unimolecular)** mechanism is exactly the same as that in an S_N1 mechanism (Section 20.2), namely, the formation of a carbocation by heterolytic fission. The E1 mechanism is therefore most likely for tertiary halogenoalkanes such as $(CH_3)_3CBr$. Figure 20.11 shows that instead of reacting with a nucleophile, the carbocation $(CH_3)_3C^+$ deprotonates.

In an E1 mechanism, once the leaving group has departed, almost anything can act as a base and remove a proton from the carbocation. Although only weakly basic, solvents such as water or alcohols are sufficient. Figure 20.12 shows that a solvent molecule capturing a proton is essentially equivalent to the proton 'just falling off' in a mechanism.

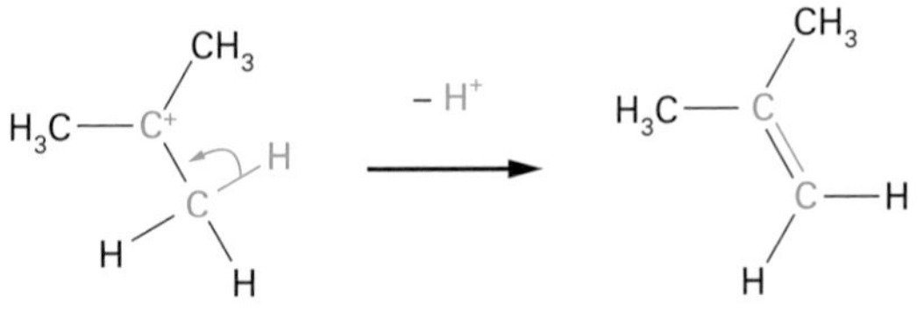

FIGURE 20.11 The carbocation $(CH_3)_3C^+$ formed during the S_N1 mechanism (Taking the Next Step 20.1) may lose a proton instead of reacting with a nucleophile. The result is an elimination reaction: Br^- is lost in the first step and H^+ in the second.

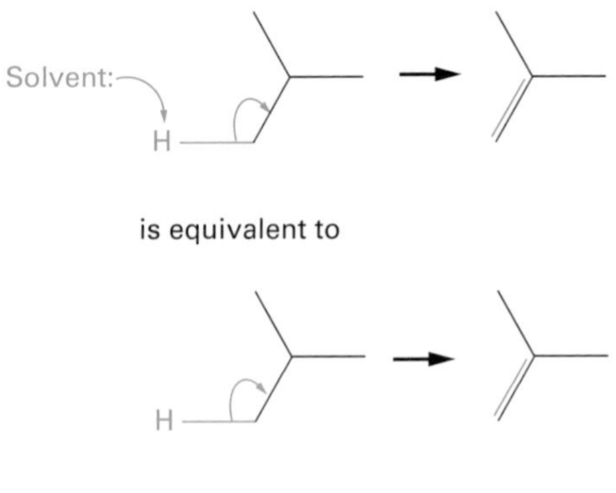

FIGURE 20.12 Often the solvent that causes deprotonation is left out and we simply show the proton 'just falling off'.

There is a third variation of the elimination mechanism, known as E1cB (cB stands for conjugate base), but this is a university topic which will not be discussed here.

20

TAKING THE NEXT STEP 20.3

What factors favour elimination over substitution?

Nucleophiles that are strong bases favour elimination over substitution. Removal of the proton during the E2 mechanism is easiest if the nucleophile is a strong base: hydroxide ion is a good example.

Nucleophiles/bases that are bulky favour elimination over substitution. A bulky group will experience steric hindrance when attacking the carbon atom in an S_N2 mechanism (Section 20.2). However, the neighbouring proton will be much easier to remove by the base. A particularly useful example is $KOC(CH_3)_3$, called potassium t-butoxide (KOt-Bu).

High temperatures favour elimination over substitution. The major factor here is that elimination always increases the number of particles, unlike substitution. Therefore, the *entropy* change is certainly positive (Section 9.1). Hence the Gibbs energy change (Section 9.3) becomes ever more negative as the temperature increases.

For the bimolecular mechanisms (S_N2 and E2), the proportion of elimination increases relative to substitution from primary to tertiary, partly as a result of the steric hindrance to S_N2 caused by more alkyl groups.

For the unimolecular mechanisms (S_N1 and E1) using hydroxide ion, each specific carbocation has a different likelihood of substituting to form the alcohol or deprotonating to form the alkene.

20

CHAPTER
21

ALCOHOLS

21.1 INTRODUCTION TO THE ALCOHOLS

- **Alcohols (ROH)** have at least one **hydroxyl group (–OH)** bonded to a carbon atom.

- We name alcohols by replacing the **–e** at the end of the corresponding alkane with **–ol**. So CH_3CH_2OH is ethanol. For longer chains, we need a number to show the position of the hydroxyl group: $CH_3CH(OH)CH_2CH_3$ is butan-2-ol.

- Alcohols undergo two types of reaction, very similar to those halogenoalkanes undergo (Sections 20.2 and 20.3): nucleophilic substitution and elimination.

University courses will also explain how alcohols react to form **ethers** (ROR').

- In addition, alcohols react with carboxylic acids to form **esters** (Section 23.4) and many alcohols also undergo **oxidation** reactions (Section 21.4).

- Taking the Next Step 21.1 describes the two main manufacturing processes for ethanol CH_3CH_2OH: fermentation and direct hydration of ethene.

    Global production of ethanol in 2017 was 27 billion US gallons.

TAKING THE NEXT STEP 21.1

The manufacture of ethanol

Fermentation uses as its starting material a sugar such as sucrose or glucose. In the presence of yeast (which contains enzymes), the absence of oxygen, and at a temperature of about 35 °C, ethanol is produced along with carbon dioxide.

Energy provided by consumption of ATP is needed.

The equation for fermentation starting from glucose is

$$C_6H_{12}O_6 \rightarrow 2CH_3CH_2OH + 2CO_2$$

At higher temperatures, the enzymes in the yeast denature and can no longer catalyse the reaction.

The **direct hydration of ethene** uses as its starting material ethene available from fractional distillation and cracking (Section 17.2) of crude oil.

Section 18.2 discusses reactions of ethene under acidic conditions.

Reaction between **ethene** and **steam** at **300 °C** and **70 atm** in the presence of a **phosphoric acid catalyst** produces ethanol:

$$C_2H_4(g) + H_2O(g) \rightarrow CH_3CH_2OH(g)$$

The temperature is a compromise. Le Chatelier's principle (Section 6.3) explains that the yield falls as temperature increases, because the reaction is exothermic (-46 kJ mol^{-1}). However, the reaction is faster at higher temperatures.

The pressure is also a compromise. Le Chatelier's principle explains that higher pressure favours the product as there is a reduction in the number of moles of gas. However, high-pressure equipment is expensive.

Fermentation has the **advantages** over direct hydration that the starting material is renewable (sugar cane can be grown) and the equipment is inexpensive. The **disadvantages** are that the process is slower, the product is much less pure (less than 15% ethanol), and manpower costs are higher.

21.2 NUCLEOPHILIC SUBSTITUTION REACTIONS OF ALCOHOLS

- Alcohols are easily made and therefore have a lot of great uses in organic synthesis.

- However, hydroxide ion HO$^-$ is a very poor leaving group so it has to be converted into a **significantly better leaving group** before any nucleophilic substitution reaction can succeed.

- We do this by using an **acid to protonate the hydroxyl group** to make the much better leaving group H$_2$O. (The presence of the acid is essential, as nucleophiles, such as Br$^-$ from KBr, on their own do *not* react.) The $-$OH$_2^+$ group also withdraws electron density from the carbon$-$oxygen bond more effectively, making the partial positive charge on carbon larger.

- One nucleophilic substitution reaction of alcohols often studied pre-university is the **reaction with potassium bromide in concentrated sulfuric acid**.

  - The alcohol ROH becomes the corresponding bromoalkane RBr.

  - Taking the Next Step 21.2 shows that the mechanism is very similar to that for the reaction of a halogenoalkane with a nucleophile (Section 20.2).

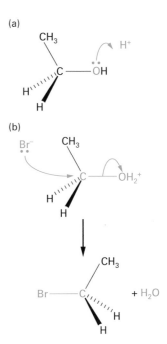

FIGURE 21.1 The mechanism for the nucleophilic substitution reaction between ethanol and potassium bromide in concentrated sulfuric acid. (a) Protonation of the alcohol's oxygen atom (which donates a lone pair to form a single covalent bond). (b) Reverse-side attack by the bromide ion nucleophile. The C$-$Br bond forms and the C$-$O bond breaks: water is the leaving group.

TAKING THE NEXT STEP 21.2

One S$_N$2 mechanism

Figure 21.1 shows the nucleophilic substitution reaction between ethanol and potassium bromide in concentrated sulfuric acid.

Notice the very close similarity (after the protonation step) to the S$_N$2 reaction of a halogenoalkane with a nucleophile (Section 20.2).

At university, you will discuss the mechanisms for conversion of the hydroxyl group into other functional groups using phosphorus reagents (such as PCl$_3$ or PCl$_5$) or sulfonate esters, but these mechanisms are beyond the scope of this book.

21.3 ELIMINATION REACTIONS OF ALCOHOLS

Phosphoric acid can be used as a catalyst for both the hydration of ethene (Section 21.1) *and* the dehydration of ethanol: this is a common feature of catalysis, as explained in detail in Section 10.2.

- As discussed previously, hydroxide ion is a very poor leaving group. Alcohols can, however, be **dehydrated by heating with acid**. Conditions depend on the specific alcohol involved, so the acid may need to be concentrated: typical reagents include concentrated sulfuric acid or concentrated phosphoric acid.

- For example, ethanol is dehydrated to ethene and 2-methylpropan-2-ol is dehydrated to 2-methylpropene:

  - $CH_3CH_2OH \rightarrow C_2H_4 + H_2O$
  - $(CH_3)_3COH \rightarrow (CH_3)_2C{=}CH_2 + H_2O$

Taking the Next Step 21.3 shows the mechanism of the 2-methylpropan-2-ol elimination reaction.

TAKING THE NEXT STEP 21.3

How does 2-methylpropan-2-ol eliminate?

As for halogenoalkanes (Section 20.3), either an E2 or an E1 mechanism can happen, once the hydroxyl group is protonated. Figure 21.2 shows the E1 elimination mechanism, which applies to tertiary alcohols such as 2-methylpropan-2-ol.

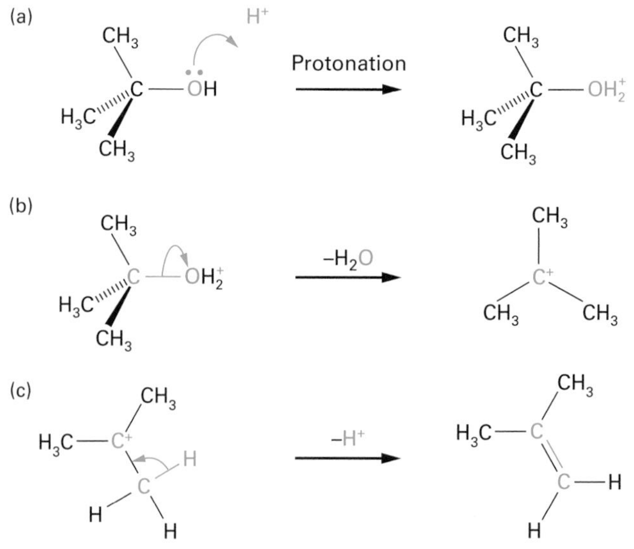

FIGURE 21.2 The mechanism for production of 2-methylpropene by elimination of water from 2-methylpropan-2-ol. (a) The oxygen atom of the tertiary alcohol, with its two lone pairs, provides a site for protonation by the catalyst, concentrated sulfuric acid. (b) The protonated alcohol loses water, which is a good leaving group. This loss creates a tertiary carbocation, which is stabilized by three electron-donating methyl groups. (c) The tertiary carbocation loses a proton, regenerating the catalyst (H^+). The product of dehydration is an alkene.

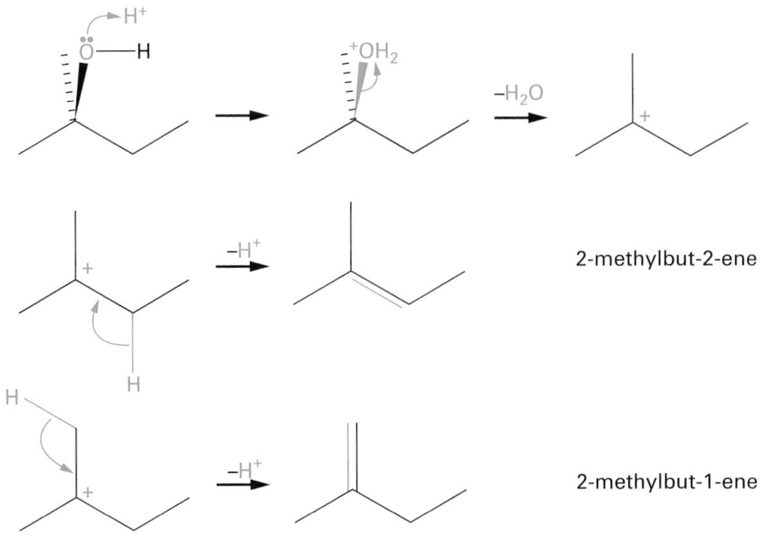

FIGURE 21.3 2-Methylbutan-2-ol can dehydrate in two ways.

- Some alcohols can dehydrate to produce **more than one product**.
  - Figure 21.3 shows that 2-methylbutan-2-ol can dehydrate to form 2-methylbut-2-ene or 2-methylbut-1-ene.

21.4 OXIDATION REACTIONS OF ALCOHOLS

- Exactly as for halogenoalkanes (Section 20.2), we classify alcohols as primary, secondary, or tertiary:
  - **Primary alcohols** have the general formula RCH_2OH: they have one alkyl group attached to the carbon bonded to the hydroxyl group.
  - **Secondary alcohols** have the general formula RR'CHOH: they have two alkyl groups attached to the carbon bonded to the hydroxyl group.
  - **Tertiary alcohols** have the general formula RR'R"COH: they have three alkyl groups attached to the carbon bonded to the hydroxyl group.
- The different classes of alcohols give similar products after nucleophilic substitution (although their mechanisms may be different). The different classes also give similar products after elimination (again, their mechanisms may be different). **The products after oxidation reactions are very different for the different classes**.
- You will probably have come across one common method for oxidizing alcohols: **heating** the alcohol with **potassium dichromate(VI)** $K_2Cr_2O_7$ acidified with **dilute sulfuric acid**.
  - This **orange** solution (with Ox(Cr) = +6) is reduced to a **green** solution (containing Cr^{3+}), thus providing a useful colour test.

When working quickly, we can write the reagent as $Cr_2O_7^{2-}/H^+$.

A Deeper Look 21.1 describes a much safer technique used at university.

21

A DEEPER LOOK 21.1

Magtrieve®

An alternative safer and greener method than using potassium dichromate(VI) is to use Magtrieve®, which is solid chromium(IV) oxide CrO_2 bound to a magnetic core. This binding to the magnetic core allows the reagent to be removed safely from the reaction, after which it is easily recycled. This avoids the use of chromium in its VI oxidation state ('hexavalent chromium'), which is toxic.

- Using the reagents described previously, we can oxidize a primary alcohol in two stages, first to an aldehyde and then to a carboxylic acid.
  - Figures 21.4 and 21.5 show that we can oxidize ethanol CH_3CH_2OH to ethanal (acetaldehyde) CH_3CHO and then to ethanoic (acetic) acid CH_3COOH.
  - A Deeper Look 21.2 explains the mechanism for oxidation of a primary alcohol to an aldehyde using acidified $K_2Cr_2O_7$.
  - Reaction does not stop at the aldehyde because the water present attacks the aldehyde to form its hydrate, which is more easily oxidized (in the same way an alcohol is: we describe the aldehyde oxidation in detail in Section 22.2).
  - To isolate the **aldehyde**, we must **distil** the product out of the reaction vessel (the aldehyde is more volatile as it lacks hydrogen bonding).
  - To maximize the yield of **carboxylic acid**, we must **reflux** the mixture, as illustrated in Figure 21.6.
- **We can oxidize a secondary alcohol to a ketone.**
  - We can oxidize propan-2-ol $CH_3CH(OH)CH_3$ to propanone (acetone) CH_3COCH_3, as shown in Figure 21.7.
  - With secondary alcohols there is no other hydrogen that can be removed, so reaction stops at the ketone.
- **Tertiary alcohols resist oxidation**, as there are no hydrogens that can be removed.

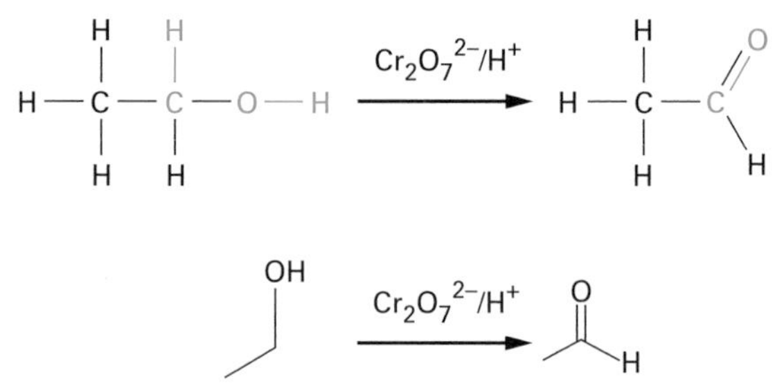

FIGURE 21.4 The oxidation of ethanol to ethanal.

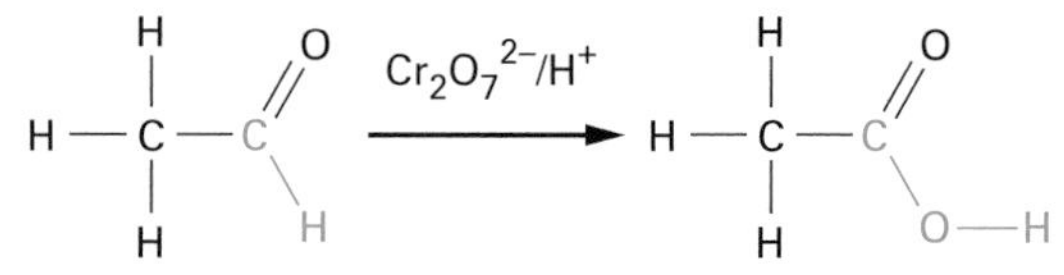

FIGURE 21.5 The oxidation of ethanal to ethanoic acid.

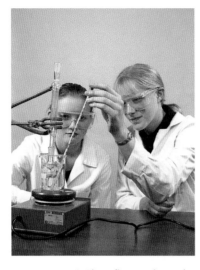

FIGURE 21.6 The reflux condenser is arranged vertically, so any vapours fall back into the flask and carry on reacting.

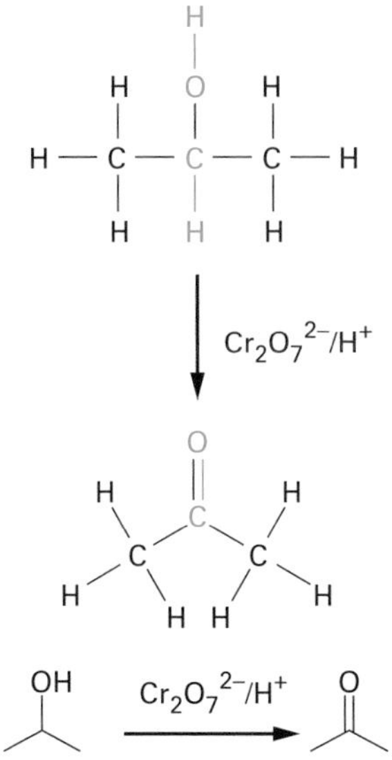

FIGURE 21.7 The oxidation of propan-2-ol to propanone.

A DEEPER LOOK 21.2

How does acidified dichromate(VI) oxidize an alcohol?

Reaction of strong acid (such as concentrated sulfuric acid) with the dichromate(VI) ion converts it into the corresponding oxide, chromium(VI) oxide (CrO_3).

Figure 21.8 shows that this oxide then forms a **chromate ester** (an inorganic ester). One hydrogen atom is removed (from the hydroxyl group) in the formation of this chromate ester. Figure 21.8 also shows the **cyclic mechanism** which removes a second hydrogen atom (from the carbon atom).

Notice how the arrows stop on the Cr atom and start again from the Cr=O bond: the net result is the addition of two electrons to the chromium. The chromium(IV) species $CrO(OH)_2$ undergoes subsequent reactions to give chromium in oxidation state III.

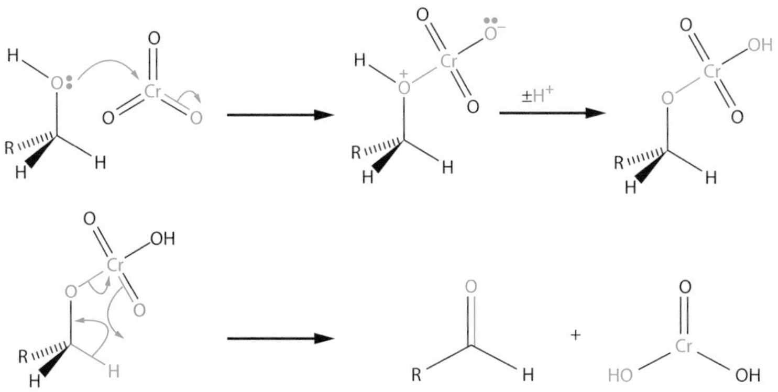

FIGURE 21.8 Oxidation of a primary alcohol to an aldehyde occurs by formation of the chromate ester, which then breaks up via a cyclic mechanism.

ALDEHYDES AND KETONES

22.1 INTRODUCTION TO THE ALDEHYDES AND KETONES

FIGURE 22.1 A carbonyl group is trigonal planar.

- Aldehydes and ketones contain the **carbonyl group**, as shown in Figure 22.1: a carbon atom doubly bonded to an oxygen atom. As expected from VSEPR (Section 2.4), the shape around the **carbonyl carbon** is trigonal planar.

- **Aldehydes (RCHO)** have one alkyl group and one hydrogen atom attached to the carbonyl carbon whereas **ketones (RCOR′)** have two alkyl groups. Figure 22.2 shows the electrostatic potential maps for ethanal (acetaldehyde), an aldehyde, and propanone (acetone), a ketone.

- To name aldehydes, we replace the **−e** at the end of the corresponding alkane with **−al**. So CH_3CHO is ethanal and CH_3CH_2CHO is propanal.

- To name ketones, we replace the **−e** at the end of the corresponding alkane with **−one**. So CH_3COCH_3 is propanone and $CH_3COCH_2CH_3$ is butanone. For longer carbon chains, we need a number to show the position of the carbonyl group: $CH_3COCH_2CH_2CH_3$ is pentan-2-one, because the oxygen is bonded to the second carbon.

(a)   (b)

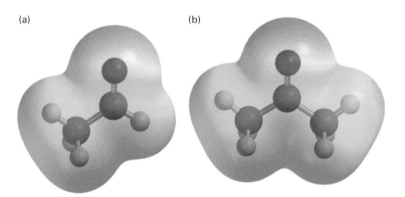

FIGURE 22.2 Electrostatic potential maps of (a) ethanal and (b) propanone show the polar nature of the carbonyl group. The oxygen atom is δ− (shown by the red colour); the carbon atom is δ+. (Remember that the Spartan software intentionally does not show the number of bonds.)

22.2 DISTINGUISHING BETWEEN ALDEHYDES AND KETONES

- **Ketones resist oxidation** (they have no hydrogens bonded to the carbonyl carbon), whereas aldehydes are relatively easily oxidized to carboxylic acids, for example by acidified potassium dichromate(VI).

  - Hence we can distinguish aldehydes from ketones by using specific mild oxidizing agents.

  - The two common tests to distinguish between aldehydes and ketones involve Fehling's solution or Tollens' reagent.

    The scientists involved are Hermann von Fehling and Bernhard Tollens, hence the position of the apostrophe for Tollens' reagent.

- We make **Fehling's solution** by adding an alkaline solution of Rochelle salt (Fehling's B) to aqueous copper(II) sulfate (Fehling's A). The blue solution contains a copper(II) complex ion.

    Rochelle salt is sodium potassium tartrate (2,3-dihydroxybutanedioate). It is the tartrate ion that complexes the copper(II) ion.

  - On warming, **aldehydes reduce the blue solution to a red-brown precipitate** (copper(I) oxide, Cu_2O), as illustrated in Figure 22.3.

  - **Ketones do not react with Fehling's solution**.

- We make **Tollens' reagent** by adding aqueous ammonia to aqueous silver nitrate made alkaline by adding sodium hydroxide. The colourless solution contains a silver complex ion, whose formula is $[Ag(NH_3)_2]^+$ (Figure 15.8).

  - On warming, **aldehydes reduce the colourless solution to a silver mirror**, as illustrated in Figure 22.4.

  - **Ketones do not react with Tollens' reagent**.

- Taking the Next Step 22.1 describes how, at university, spectroscopy is used to distinguish between aldehydes and ketones.

A Deeper Look 22.1 shows the mechanism of the oxidation of (the hydrate of) an aldehyde by acidified dichromate(VI).

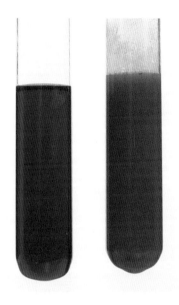

FIGURE 22.3 Aldehydes reduce Fehling's solution from a blue solution to a red-brown precipitate.

FIGURE 22.4 Aldehydes reduce Tollens' reagent to form a silver mirror.

TAKING THE NEXT STEP 22.1

How can spectroscopy distinguish aldehydes from ketones?

Carbon-13 NMR is of little use for distinguishing between aldehydes and ketones (aldehydes have typical δ values of 195 to 205 ppm, whereas ketones have typical values of 195 to 215 ppm). However, ***proton*** (^{1}H) NMR is very helpful as the characteristic signal from a **−CHO** group (which is absent in a ketone) has a δ value of **9 to 10 ppm**, as shown in Figure 22.5.

22

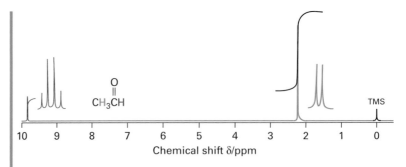

FIGURE 22.5 The ^{1}H NMR spectrum of ethanal (acetaldehyde) CH_3CHO. Notice the splitting (Section 26.6): the CH_3 peak just above 2 ppm is a 1:1 doublet and the CH peak between 9 and 10 is a 1:3:3:1 quartet.

Simple aldehydes absorb infrared radiation at a slightly higher frequency than simple ketones (1730 cm^{-1} compared with 1715 cm^{-1}): remember, hydrogen is the lightest atom (Figure 11.1). The C–H bond (Section 26.2) also shows up at 2700 to 2770 cm^{-1}, as seen in Figure 22.6, often as a doublet (Figure 26.5).

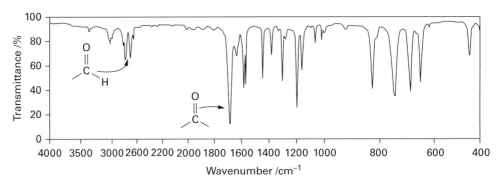

FIGURE 22.6 The infrared spectrum of benzaldehyde, C_6H_5CHO.

A DEEPER LOOK 22.1

Oxidation of an aldehyde

An aqueous aldehyde is in equilibrium with its hydrate, formed as shown in Figure 22.7.

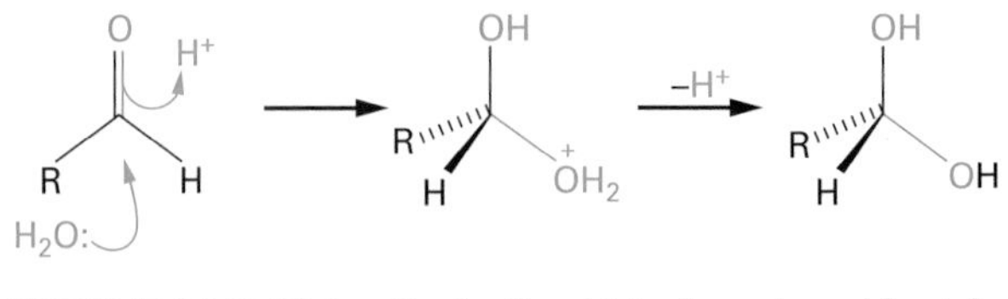

FIGURE 22.7 Water attacks the aldehyde, under acid catalysis, to form the hydrate.

The hydrate is then oxidized in the same way as an alcohol is (Figure 21.8), as shown in Figure 22.8.

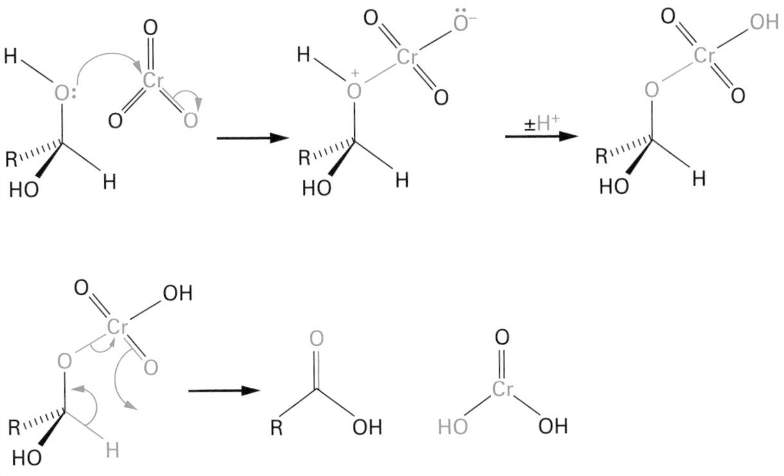

FIGURE 22.8 The aldehyde's hydrate is oxidized in the same way as an alcohol is (Figure 21.8).

22.3 REDUCTION OF ALDEHYDES AND KETONES

- We can form aldehydes and ketones by oxidizing primary and secondary alcohols respectively (Section 21.4). Hence it is reasonable to predict that:

  - **We can reduce aldehydes and ketones to primary and secondary alcohols respectively**.

    ○ The carbonyl group C=O becomes CH(OH), e.g. CH_3CHO becomes CH_3CH_2OH.

- One reagent often used (in aqueous or alcoholic solution) for the reduction of aldehydes and ketones is sodium tetrahydridoborate (sodium borohydride), **NaBH$_4$**.

- Figure 22.9 shows the **mechanism for the reduction of aldehydes and ketones** you are likely to have come across pre-university, where the reagent has been greatly simplified to a **hydride ion**, H⁻: see Taking the Next Step 22.2. (Hydride ion itself is too strong a base to be used.)

An alternative reagent is the more powerful reducing agent lithium tetrahydridoaluminate (lithium aluminium hydride, **lithal**), **LiAlH$_4$**. LiAlH$_4$ requires a dry solvent (typically an ether such as ethoxyethane), since it reacts with water.

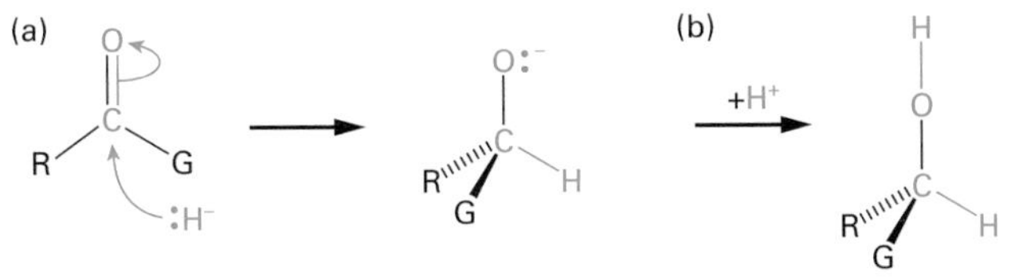

FIGURE 22.9 Reduction of a carbonyl group by NaBH$_4$. G can be R' or H. (a) A hydride ion H⁻ carries out nucleophilic attack on the δ+ carbonyl carbon atom. (b) Protonation by the solvent forms the product.

What is the mechanism for reduction using borohydride ion?

Figure 22.10 is a *much* better representation than the hydride ion transfer: the boron–hydrogen bond provides the electron pair.

The solvent (typically water) provides the proton needed to form the alcohol (ROH) from the alkoxide (RO⁻).

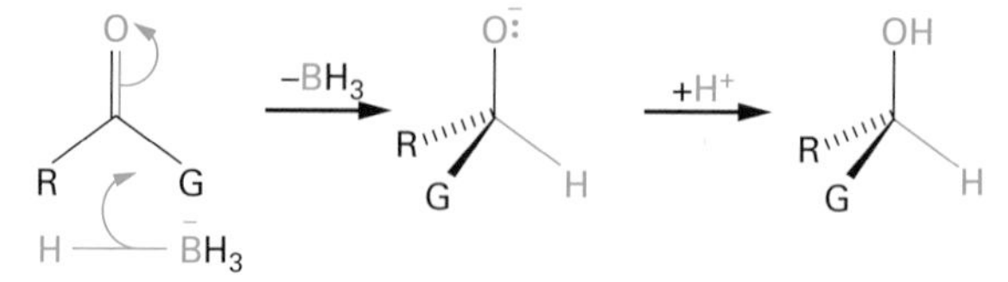

FIGURE 22.10 The electron pair causing the reduction comes from the boron–hydrogen bond.

22.4 NUCLEOPHILIC ADDITION TO ALDEHYDES AND KETONES

A Deeper Look 22.1 shows the similar mechanism for the formation of a hydrate, using water as the nucleophile (although protonation occurs first in the case of hydrate formation).

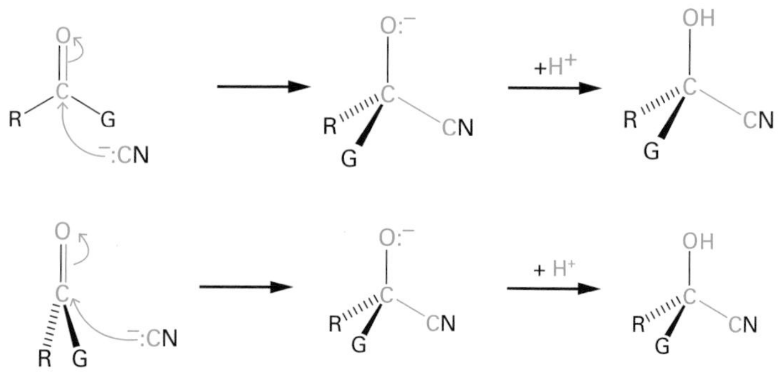

- The simplified mechanism of reduction shown in Figure 22.9 involves attack by the nucleophilic hydride ion at the partially positive carbon atom, followed by protonation.

- Another **nucleophilic addition** reaction is that of **HCN** (formed from potassium cyanide and strong acid) which also involves attack by the nucleophilic cyanide ion at the partially positive carbon atom, followed by protonation.

- Figure 22.11 shows the **mechanism for the nucleophilic addition of HCN** to a carbonyl group.
  - In this reaction, the carbonyl group C=O becomes a **hydroxynitrile (cyanohydrin)** COH(CN).
  - See Figure 16.14 for a discussion of the **stereochemistry** of this reaction.

- One particularly useful nucleophilic addition reaction in organic synthesis involves the **Grignard reaction**: see Taking the Next Step 22.3. The Grignard reaction provides an excellent way of making a new carbon–carbon bond.

FIGURE 22.11 Nucleophilic addition to a carbonyl group leads to an alcohol with a carbon–nucleophile single bond. Note that there is no other product. The view of the addition of cyanide in the second line gives a more accurate picture of the attack of the ion. G may be R' or H.

22

TAKING THE NEXT STEP 22.3

What is the Grignard reaction?

The **Grignard reaction** involves a nucleophilic addition reaction, as shown in Figure 22.12, between a Grignard reagent and an aldehyde or ketone, making an alcohol.

We can make a **Grignard reagent** by dissolving a halogenoalkane in a dry solvent (normally an ether; either ether (ethoxyethane) itself or THF, Figure 22.13) and reacting with magnesium turnings. It is usual to beat the turnings first with a stirrer bar as magnesium tends to have a layer of magnesium oxide on it (because air oxidizes the surface). The resulting reagent, as shown in Figure 22.14, has the simplified formula of RMgX, where X is a halogen.

To simplify the mechanism, as shown in Figure 22.15, we can consider the Grignard reagent as providing a **carbanion** R^-, which is an ion in which carbon carries a negative charge (making it a popular carbon nucleophile). The bond between the carbon atom and the magnesium is highly polarized towards the more electronegative carbon atom.

(a)

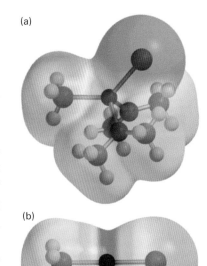

(b)

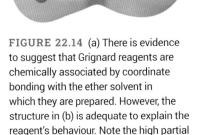

FIGURE 22.14 (a) There is evidence to suggest that Grignard reagents are chemically associated by coordinate bonding with the ether solvent in which they are prepared. However, the structure in (b) is adequate to explain the reagent's behaviour. Note the high partial positive charge on the magnesium atom, shown by the blue colour.

(a)

(i) CH_3CH_2MgI
(ii) H^+ from acid

→ CH_3CH_2—C—OH Propan–1–ol

(b)

(i) CH_3MgI
(ii) H^+ from acid

→ CH_3—C—OH Propan–2–ol

(c)

(i) CH_3MgI
(ii) H^+ from acid

→ CH_3—C—OH 2–Methylpropan–2–ol

(d)

O=C=O

(i) CH_3CH_2MgI
(ii) H^+ from acid

→ CH_3CH_2—C Propanoic acid

FIGURE 22.12 A Grignard reagent adds to an aldehyde or ketone to produce an alcohol. (d) Shows the reaction between a Grignard reagent and carbon dioxide.

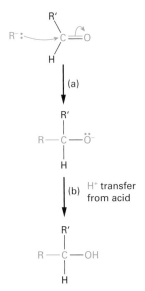

FIGURE 22.15 (a) An addition reaction of a Grignard reagent. (b) Acid added during work-up causes protonation of the addition product.

FIGURE 22.13 The solvent tetrahydrofuran (THF).

22

22.5 CONDENSATION REACTIONS OF ALDEHYDES AND KETONES

- A number of nucleophilic addition reactions of aldehydes and ketones are followed by a subsequent elimination reaction; we then call the overall process a **condensation reaction**.

- Taking the Next Step 22.4 shows the mechanism for a general condensation reaction of nitrogen nucleophiles with aldehydes and ketones.

TAKING THE NEXT STEP 22.4

What is the mechanism for condensation reactions?

Some nitrogen nucleophiles, listed in Table 22.1, undergo elimination after the nucleophilic addition reaction, as shown in Figure 22.16, to give an overall condensation reaction.

We can use one such condensation reaction to identify aldehydes and ketones, because the products of reaction with **2,4-dinitrophenylhydrazine (DNP)** (see A Deeper Look 22.2) are yellow/orange solids, which can be easily purified.

The melting point of the DNP derivative was used in the past to identify the original carbonyl compound, before NMR analysis became routine.

> Safety organizations have consistently stated the need to ensure that a DNP solution should not be allowed to dry out as it can then become explosive. In 2016 at least ten schools, concerned about the chance that a sample had dried out, called in bomb disposal squads to destroy the sample with a controlled explosion on the playing fields (occasionally forgetting to warn the neighbours!).

A Deeper Look 22.2 also discusses other similar nitrogen nucleophiles.

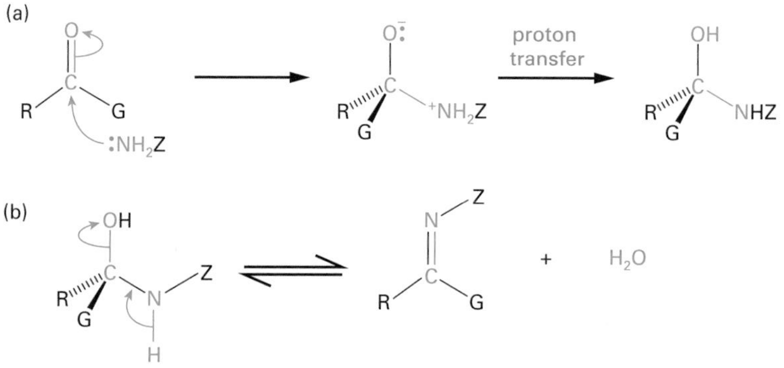

FIGURE 22.16 (a) Nucleophilic addition and (b) condensation. G may be R' or H. When Z is an alkyl group, we call the product formed in a condensation reaction an **imine**. Biochemists often call an imine a **Schiff base**: see A Deeper Look 22.3.

A DEEPER LOOK 22.2

What other nucleophiles give condensation reactions?

Table 22.1 shows examples of nitrogen nucleophiles that produce a condensation product.

TABLE 22.1 Examples of nucleophiles that produce a condensation product. The product shown is that formed from propanone (acetone). The right-hand column shows the name given to the class of product.

Nucleophile	Product	Class of Product
H_2N—R	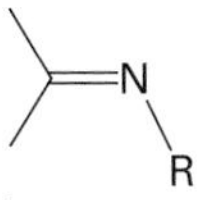 (structure)	imine
H_2N—OH	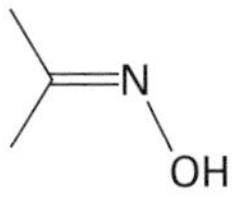 (structure)	oxime
H_2N—NH_2	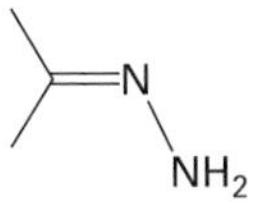 (structure)	hydrazone
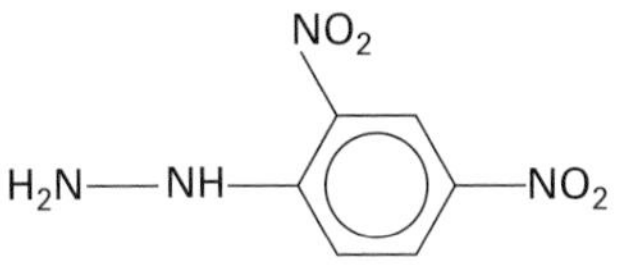 H_2N—NH—(2,4-dinitrophenyl)	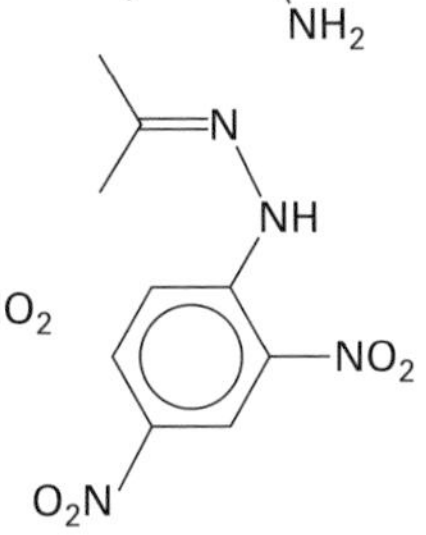 (structure)	2,4-dinitrophenylhydrazone

A DEEPER LOOK 22.3

Nature uses a Schiff base to form alanine

Nature's synthesis of the amino acid alanine (Table 24.1) starts with reaction between the amine pyridoxamine and the carbonyl compound pyruvic acid (2-oxopropanoic acid), Figure 22.17, to form a Schiff base. This Schiff base forms an equilibrium mixture with a different Schiff base. The latter can then be hydrolysed to form the aldehyde pyridoxal and alanine.

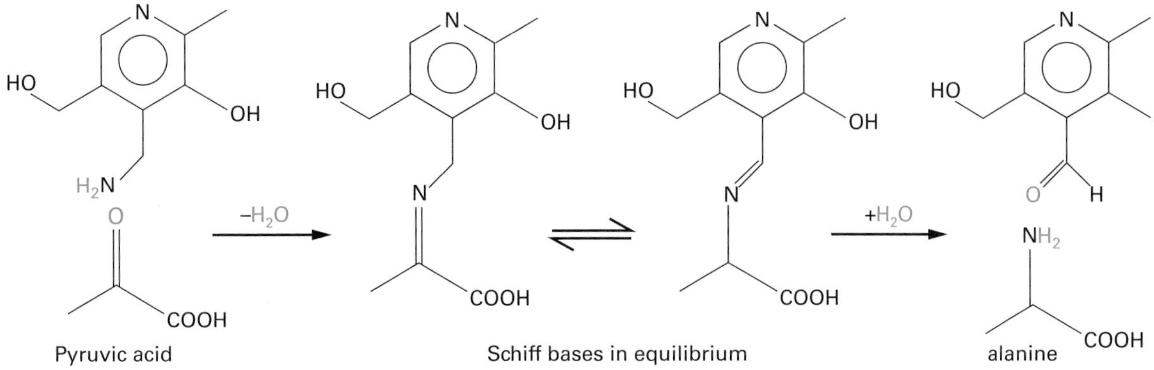

FIGURE 22.17 Nature's synthesis of the amino acid alanine.

22

22.6 REACTION AT THE ALPHA CARBON

One final type of reaction that aldehydes and ketones undergo, which is particularly important at university, involves the removal of a proton from the carbon atom immediately *next* to the carbonyl carbon, called the **alpha carbon**. A Deeper Look 22.4 describes the **aldol reaction**.

A DEEPER LOOK 22.4

What are enols, enolates, and the aldol reaction?

Section 10.3 showed that plotting the concentration of a reactant against time gives a straight line for a *zero*-order reaction. Figure 22.18 shows the graph of iodine concentration against time for the acid-catalysed reaction between propanone (acetone) and iodine

$$CH_3COCH_3 + I_2 \rightarrow CH_3COCH_2I + HI$$

We can interpret this as a slow step *not involving iodine* and a fast one that does involve iodine, as shown in Figure 22.19.

(a) Initial concentrations

iodine	0.01 mol dm^{-3}
propanone	0.25 mol dm^{-3}
sulfuric acid	0.25 mol dm^{-3}

At $t = 0$, $[I_2] = 0.01$ mol dm^{-3}

(b) [graph of $[I_2]$ / mol dm^{-3} against Time / min]

(c)
$$\text{rate} = \frac{\text{change in conc.}}{\text{time}}$$
$$= \frac{0.003 \text{ mol dm}^{-3}}{30 \text{ min}}$$
$$= 1 \times 10^{-4} \text{ mol dm}^{-3} \text{ min}^{-1}$$

[graph of Rate / mol dm^{-3} min^{-1} against $[I_2]$ / mol dm^{-3}]

FIGURE 22.18 The graph of iodine concentration against time is a straight line, which shows that the reaction is zero-order with respect to iodine (Section 10.3).

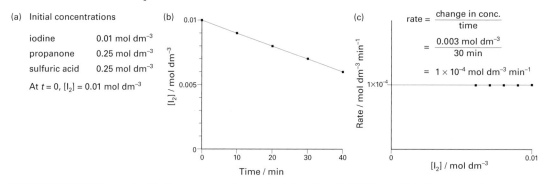

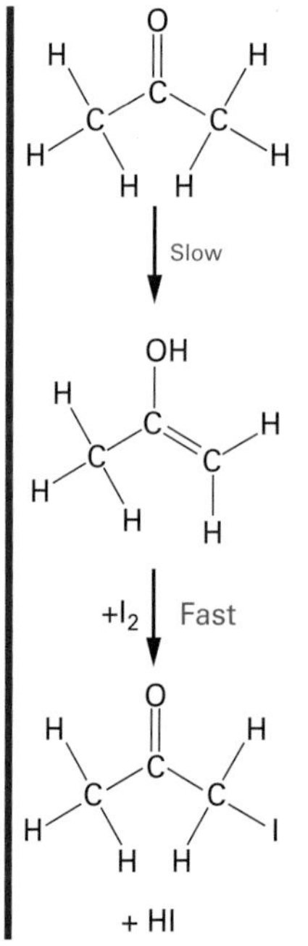

FIGURE 22.19 The rate-limiting keto–enol tautomerism is followed by rapid reaction with iodine.

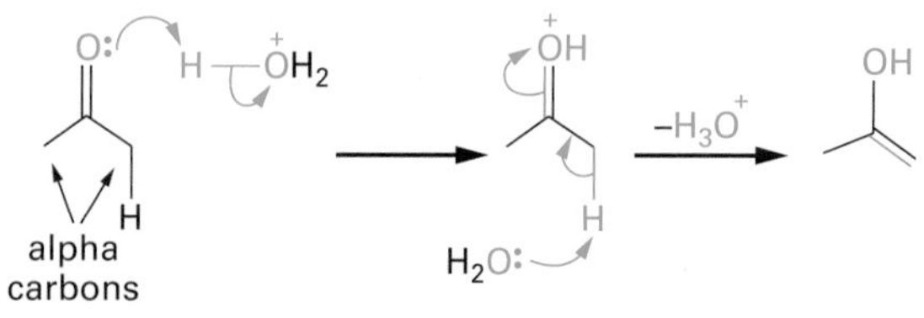

alpha carbons

FIGURE 22.20 Keto-enol tautomerism: addition of a proton onto oxygen followed by removal of a proton from carbon forms an enol.

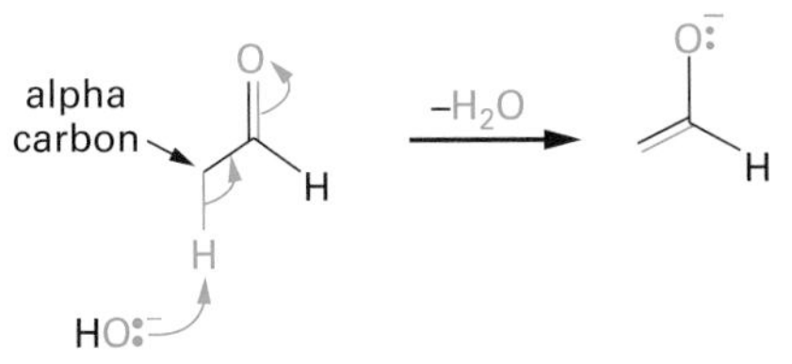

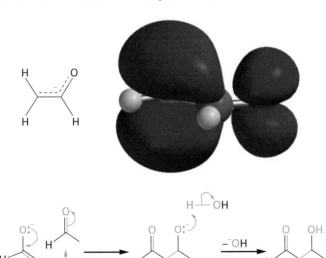

FIGURE 22.21 Removal of a proton from carbon by a hydroxide ion forms an enolate ion.

Let us think again about the slow step. The acid present protonates the oxygen atom and then a water molecule causes deprotonation from one of the two neighbouring carbon atoms (each is an alpha carbon), as shown in Figure 22.20. We call the resulting molecule an **enol** because of the double bond (en) and the hydroxyl group (ol). (We call the process **keto–enol tautomerism**.) The double bond then reacts quickly with iodine (Section 18.4).

For enol formation there must be at least one alpha carbon with at least one hydrogen atom attached to it. So ethanal (acetaldehyde) CH_3CHO can also form an enol. A very similar reaction occurs in basic solution where hydroxide ion removes a proton, as shown in Figure 22.21 (from ethanal in this case), to form an **enolate ion**.

Figure 22.22 shows that the negative charge in the enolate ion from ethanal is delocalized with the double bond, which is what makes the molecule deprotonate in the first place. Calculations on the HOMO (A Deeper Look 16.3) of the enolate ion show that the highest electron density is on the carbon.

Formation of the enolate has one other consequence. The enolate is a nucleophile and so can react by nucleophilic addition with the original unenolized molecule, as shown in Figure 22.23 (using ethanal as an example). We call the product (on protonation) an **aldol**, as it is an *ald*ehyde and an alco*hol*.

The aldol reaction is very important in organic synthesis as it makes a new carbon–carbon bond. For example, an aldol reaction is used in the

FIGURE 22.22 The enolate ion is delocalized. The HOMO of the enolate ion shows electron density on both the carbon and the oxygen, with the highest electron density on the carbon.

FIGURE 22.23 An aldol reaction occurs by reaction between an enolate ion and the original unenolized molecule.

preparation of the drug atorvastatin (Lipitor™) used to treat cardiovascular disease, which in 2019 was the third most prescribed drug in the US.

In a **crossed aldol reaction**, the enolate from one carbonyl compound reacts with the carbonyl group of a different carbonyl compound. A biochemical example occurs in the **Krebs cycle** (or **citric acid cycle**), which is an important metabolic pathway for the oxidation of fuel molecules (such as carbohydrates). Most fuel molecules enter the cycle as acetyl coenzyme A: 'acetyl' is the traditional name for 'ethanoyl'. The first step of the Krebs cycle, Figure 22.24, involves a crossed aldol reaction between acetyl coenzyme A ($CH_3COSCoA$) and the four-carbon oxaloacetate ion, catalysed by the enzyme citrate synthase. After hydrolysis, the net effect is the addition of two carbons to form the six-carbon citrate ion.

In some cases, the aldol product can react further by dehydration to give a molecule in which the newly formed C=C bond is delocalized with the C=O bond. We then call such a reaction an **aldol condensation**. For example, propanone and benzaldehyde (C_6H_5CHO) react to give a good yield of $C_6H_5CH=CHCOCH_3$. (Benzaldehyde has no alpha carbon atoms, which avoids competing reactions.)

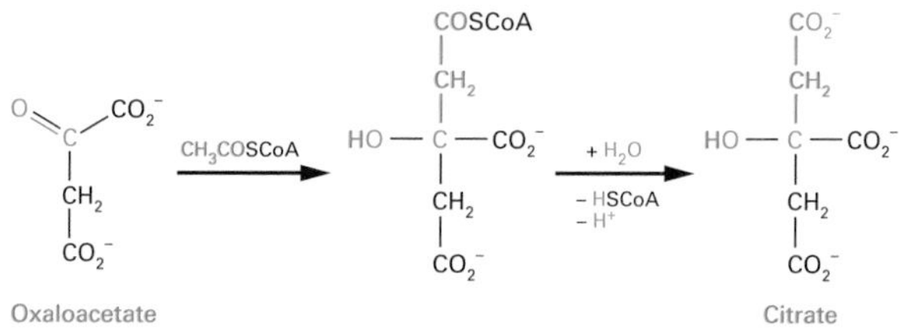

FIGURE 22.24 The first step in the Krebs cycle involves a crossed aldol reaction.

CARBOXYLIC ACIDS AND THEIR DERIVATIVES

23.1 INTRODUCTION TO THE CARBOXYLIC ACIDS

- **Carboxylic acids** have the general formula **RCOOH** and are weak acids (see Taking the Next Step 23.1), which therefore neutralize alkalis and react with carbonates (to evolve CO_2: $2H^+ + CO_3^{2-} \rightarrow CO_2 + H_2O$). They also perform redox reactions (Section 8.3) with reactive metals to give hydrogen gas.

- We name carboxylic acids by replacing the **−e** at the end of the corresponding alkane with **−oic acid**. So CH_3COOH is **ethanoic acid** (which is also called acetic acid, Section 16.2).

- We can make carboxylic acids by the oxidation of primary alcohols (Section 21.4) or aldehydes (Section 22.2) or by hydrolysis of derivatives such as esters or acyl chlorides (Sections 23.3 and 23.4).

We can also make a carboxylic acid by reacting a Grignard reagent (Section 22.4) with carbon dioxide.

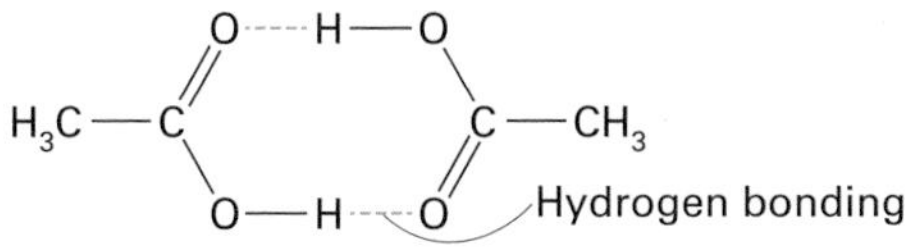

- Carboxylic acids commonly form **dimers** in the solid state because of **hydrogen bonding**, as shown in Figure 23.1. This means that carboxylic acids have higher melting points than might be expected: hexanoic acid melts at −3 °C, whereas hexan-1-ol melts at −45 °C. Hydrogen bonding (along with the acidity) also explains the good water solubility of the acids.

FIGURE 23.1 Two hydrogen bonds form between molecules of ethanoic acid, holding the molecules together in a dimer.

TAKING THE NEXT STEP 23.1

Why are carboxylic acids acidic?

Writing the equilibrium equations for the deprotonation of ethanol and ethanoic (acetic) acid suggests they are very similar:

$$CH_3CH_2OH(aq) + H_2O(l) \rightleftharpoons H_3O^+(aq) + CH_3CH_2O^-(aq)$$

$$CH_3COOH(aq) + H_2O(l) \rightleftharpoons H_3O^+(aq) + CH_3COO^-(aq)$$

but the latter equation fails to take account of the delocalization in the ethanoate (acetate) ion (Section 2.3). Figure 23.2(b) shows its electrostatic potential map, which confirms that the two oxygen atoms in the ethanoate ion are equivalent.

This delocalization in the anion, which is better written as $CH_3CO_2^-$ rather than CH_3COO^- as the oxygens are indistinguishable, stabilizes the ion and so makes deprotonation easier, which makes ethanoic acid *much* more acidic than ethanol. The pK_a values (Section 7.3) are 4.8 for ethanoic acid and 16 for ethanol: the difference is eleven orders of magnitude. At pH 4.8, half (Section 7.5) of the ethanoic acid would be deprotonated, while almost all of the ethanol would remain un-ionized.

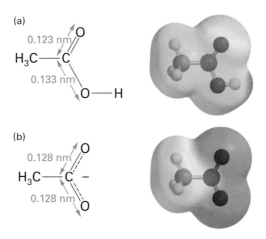

FIGURE 23.2 Electrostatic potential maps. (a) The un-ionized ethanoic acid molecule: the C–O bond is longer than the C=O bond. (b) The ethanoate ion: the negative charge in the ethanoate ion is delocalized. Both C–O bonds are equivalent, having the same bond lengths and electron densities (intermediate between C–O and C=O).

23.2 ACID DERIVATIVES

- **Carboxylic acid derivatives** differ from aldehydes and ketones in that the C=O group is attached to *an electronegative atom with at least one lone pair*. This allows for a different mechanism to that shown by aldehydes and ketones. Now the nucleophile can **replace** the electronegative atom which acts as a leaving group, as shown in Figure 23.3, as elimination can follow nucleophilic addition.

- Taking the Next Step 23.2 shows a generic mechanism for the reaction between a carboxylic acid derivative and a nucleophile. **LG** stands for the leaving group.

TAKING THE NEXT STEP 23.2

Nucleophilic addition-elimination reactions of acid derivatives

Figure 23.3 shows how a nucleophile can react with a carboxylic acid derivative.

The intermediate formed in Figure 23.3 has a carbon atom that forms four bonds. As a result, it is tetrahedral and so the (not very inventive) name used is the **tetrahedral intermediate**.

Two factors are important for deciding how reactive a particular acid derivative is. First, the lone pair on LG can overlap and **partially delocalize** with the carbonyl group, as shown in Figure 23.4. We can judge the extent of this interaction using IR stretching frequencies; see A Deeper Look 23.1. The strongest interaction is in amides, as in Figure 23.4, and the weakest is in acyl chlorides.

The C=O bond will reform (this is thermodynamically favourable as it is very strong) *if* there is a good leaving group. So the second factor is **how good the LG⁻ ion is as a leaving group**. This can be judged by the acidity of its conjugate acid (Section 7.1) HLG. The best leaving group will be a halide ion such as Cl⁻ or Br⁻ as HCl and HBr are strong acids (Section 7.2). RO⁻ will be much less good as a leaving group as alcohols have pK_a values of about 16.

Putting both the factors together, **acyl chlorides will react much more easily than carboxylic acids or esters. Least reactive will be acid amides** (Section 24.6).

One example of this general mechanism is the reaction of an acyl chloride with an alcohol, shown in Section 23.3.

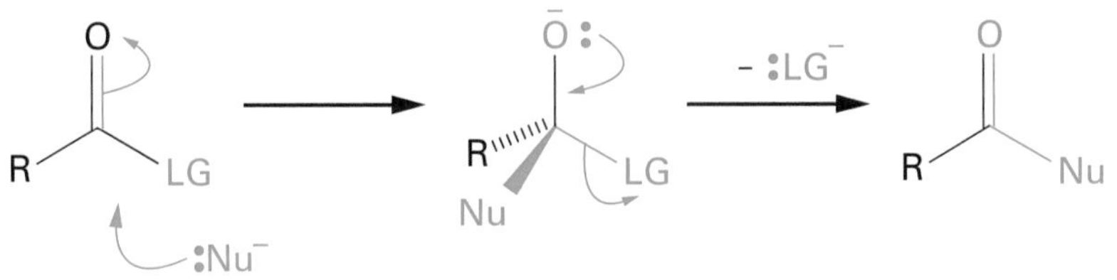

University textbooks often call this **nucleophilic acyl substitution**.

FIGURE 23.3 Nucleophilic addition is followed by elimination of the anion LG⁻.

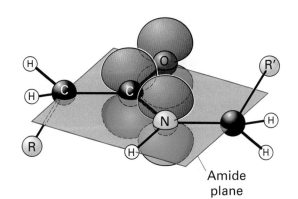

FIGURE 23.4 The p orbitals on carbon and oxygen that form the π bond can also overlap with the p orbital on nitrogen, when this is arranged as in the figure to cause a structure delocalized over all three atoms. This delocalization is the reason why the peptide bond (Section 24.6) is planar.

Amide plane

23

Using IR stretching frequencies

The strongest interaction (partial delocalization) between the lone pair on the leaving group and the carbonyl group occurs with nitrogen in an amide (C=O stretch below 1700 cm^{-1}): see Section 23.3. There is an intermediate level of interaction with oxygen: an ester has a C=O stretch between about 1735 and 1780 cm^{-1}. The least interaction is in an acyl chloride (C=O stretch about 1800 cm^{-1}) as chlorine's lone pair is in a 3p orbital.

23.3 ACYL CHLORIDES

- We can make **acyl chlorides** (also frequently called **acid chlorides**), with general formula **RCOCl** (as shown in Figure 23.5(a)), from the corresponding carboxylic acid RCOOH using a chlorinating agent. There must be no water present, as the acyl chloride would then be hydrolysed back to the original acid.
  - **Chlorinating agents** include phosphorus pentachloride PCl_5, phosphorus trichloride PCl_3, and thionyl chloride $SOCl_2$.

- We name an acyl chloride by changing the ending of the acid from which it is formed from '**–oic acid**' to '**–oyl chloride**', so ethanoic (acetic) acid becomes **ethanoyl (acetyl) chloride**.

- The typical reactions of acyl chlorides are with nucleophiles such as water, alcohols, ammonia, and amines:
  - $RCOCl + H_2O \rightarrow RCOOH + HCl$
  - $RCOCl + R'OH \rightarrow RCOOR' + HCl$
  - $RCOCl + 2NH_3 \rightarrow RCONH_2 + NH_4^+Cl^-$
  - $RCOCl + 2R'NH_2 \rightarrow RCONHR' + R'NH_3^+Cl^-$

- $RCONH_2$ is an **amide**; RCONHR' is a substituted amide.

- We call these **acylation** reactions because in each case an **acyl group** (**RCO–**) replaces a hydrogen atom in the attacking nucleophile.

- Taking the Next Step 23.3 shows the mechanism of an acylation reaction of an alcohol using an acyl chloride.

- A Deeper Look 23.2 describes how we can also do acylation reactions using acid anhydrides (Figure 23.5(b)).

We can make acyl bromides with the corresponding brominating agents.

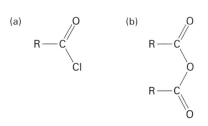

FIGURE 23.5 (a) An acyl chloride. (b) An acid anhydride.

TAKING THE NEXT STEP 23.3

One nucleophilic addition–elimination mechanism

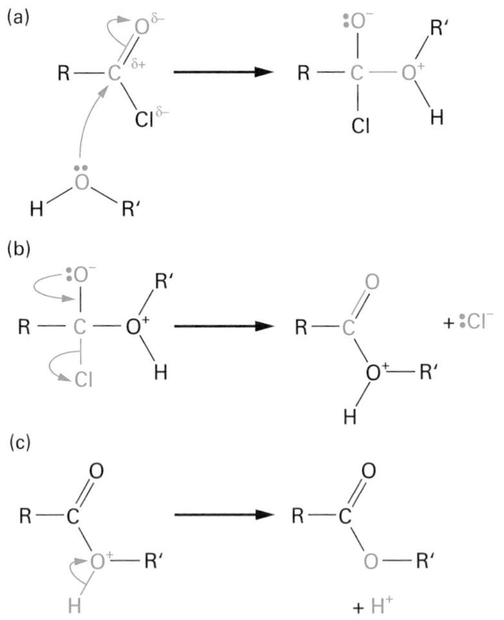

FIGURE 23.6 The **nucleophilic addition–elimination** mechanism of acylation occurs by nucleophilic addition of the alcohol followed by elimination of HCl: (a) nucleophilic attack at the carbonyl carbon atom to form a **tetrahedral intermediate**; (b) loss of a chloride ion as the double bond reforms; (c) loss of a proton.

Some pre-university textbooks (and mark schemes) show the last two steps as occurring simultaneously, which is *very* unlikely. However, deprotonation may in some cases occur before the loss of the chloride ion as a protonated alcohol is a strong acid.

A DEEPER LOOK 23.2

Acid anhydrides

We can make **acid anhydrides** (Figure 23.5(b)), with general formula **$(RCO)_2O$**, by reacting the salt (often the sodium salt) of a carboxylic acid with an acyl chloride. As for the synthesis of acyl chlorides, to avoid hydrolysis back to the carboxylic acid there must be no water present.

We name an acid anhydride by changing the second part of the name of the acid from which it is derived from '**acid**' to '**anhydride**' so ethanoic (acetic) acid becomes **ethanoic (acetic) anhydride**.

The anhydride could be *unsymmetrical*, in which case we name the two acids in alphabetical order and add 'anhydride' at the end.

Acid anhydrides react in a very similar way to acyl chlorides, albeit rather **more slowly** (as RCO_2^- is formed and carboxylic acids are weak acids with pK_a values of about 4 or 5 rather than being strong acids like HCl). The second product is RCOOH rather than HCl.

This has advantages in the manufacture of aspirin, as the less vigorous reaction of the anhydride is more easily controlled.

23.4 ESTERS

The role of the acid catalyst is to protonate the oxygen to make the carbonyl carbon more susceptible to nucleophilic attack.

- We can make **esters**, with general formula **RCOOR'**, from the corresponding carboxylic acid RCOOH and alcohol R'OH in the presence of a concentrated sulfuric acid catalyst. This forms an equilibrium mixture (Section 6.1) since there is always a choice as to which group leaves the tetrahedral intermediate (Section 23.2).

- We name an ester by following the name of the alcohol (converted into the corresponding alkyl group) with the name of the acid (with the ending changed to '−oate'). So ethanol CH_3CH_2OH and ethanoic (acetic) acid CH_3COOH react to form **ethyl ethanoate** (acetate) $CH_3COOCH_2CH_3$. Methanol CH_3OH and propanoic acid CH_3CH_2COOH react to form methyl propanoate $CH_3CH_2COOCH_3$.

- We can also make esters from an **acyl chloride RCOCl** and an **alcohol R'OH**.

  - The use of the acyl chloride is preferable as the equilibrium lies much more in favour of the ester when compared with the reaction involving the carboxylic acid, because chloride ion is a very good leaving group (Section 20.1).

    Esters are used as
    - solvents (such as ethyl ethanoate (acetate) in modelling glues),
    - plasticizers for polymers (commonly dioctyl phthalate), and
    - food additives (pear drop essence is 2-methylbutyl ethanoate, which is also called amyl acetate).

- Esters can be **hydrolysed** by acids (which produces an equilibrium mixture). However, alkalis work much better, as the alkali neutralizes the acid formed and so shifts the equilibrium in favour of the products. Taking the Next Step 23.4 shows one common mechanism for ester hydrolysis.

  - $CH_3COOCH_2CH_3 + NaOH \rightarrow CH_3CH_2OH + CH_3CO_2Na$

    When applied to oils and fats (which are triesters), alkaline hydrolysis with sodium or potassium hydroxide produces the molecule **glycerol** (propane-1,2,3-triol) and the **sodium or potassium salts of the fatty acids** present in the oil or fat (which can act as **soaps**). We call alkaline hydrolysis of an ester **saponification**, because it is the process used to make soap (Latin *sapo*).

TAKING THE NEXT STEP 23.4

Ester hydrolysis

Ester hydrolysis occurs by a number of different mechanisms, which you will study extensively at university. Figure 23.7 shows one significant mechanism seen when hydrolysis occurs in base/alkali; we call this mechanism

$B_{AC}2$ (A Deeper Look 23.3 explains the notation and also discusses other mechanisms). Figure 23.8 shows a shortened version of this mechanism which you might encounter at university. A Deeper Look 23.4 describes evidence for the tetrahedral intermediate.

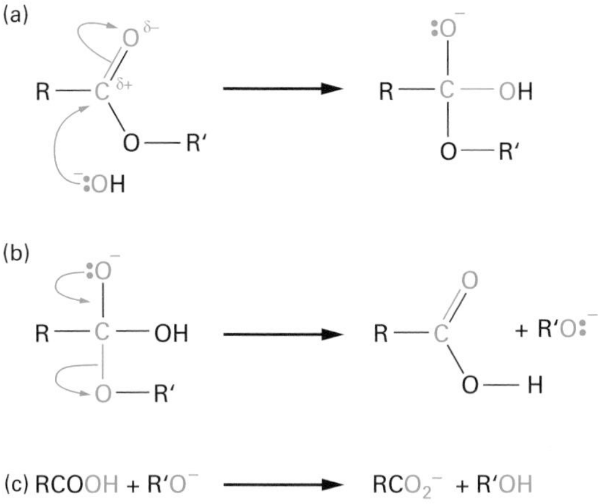

FIGURE 23.7 The $B_{AC}2$ mechanism of ester hydrolysis. (a) Nucleophilic attack by hydroxide ion on the carbon atom forming a tetrahedral intermediate. (b) Breakdown of the tetrahedral intermediate. (c) Proton transfer from carboxylic acid to alkoxide ion.

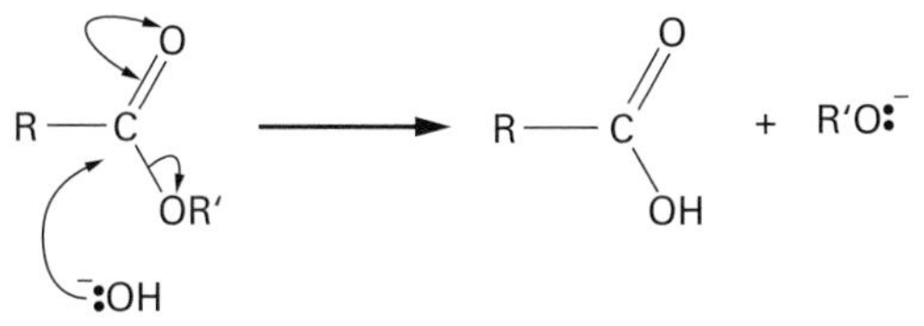

FIGURE 23.8 A shortened version, of steps (a) and (b) above, of $B_{AC}2$ ester hydrolysis.

A DEEPER LOOK 23.3

Are there other ester hydrolysis mechanisms?

The notation $B_{AC}2$ used in Taking the Next Step 23.4 stands for base catalysis, acyl fission, bimolecular. Ester hydrolysis can occur under either **acid** or **base catalysis**. **Acyl (AC) fission** means that the ester breaks RCO−OR′: the ester may break by **alkyl (AL) fission** RCOO−R′. Finally, the mechanism could be **unimolecular** rather than **bimolecular**. This would suggest that 2×2×2 = 8 different mechanisms should exist. However, we have only found six so far ($B_{AC}1$ and $A_{AL}2$ have not been seen).

In $A_{AC}2$, for example, the acid protonates the ester first, which is then attacked by water. This hydrolysis mechanism is the backward reaction of the normal ester formation mechanism, in which sulfuric acid protonates the carboxylic acid, which is then attacked by the alcohol.

In $A_{AL}1$ the ester breaks down, without the involvement of the catalyst, into RCO_2^- and R'^+, which will be most likely for esters that would then generate stable carbocations, especially tertiary carbocations (Section 18.3). $A_{AL}1$ occurs for *t*-butyl ethanoate (acetate) in acid as the *t*-butyl carbocation $(CH_3)_3C^+$ is formed.

23

A DEEPER LOOK 23.4

Myron Bender's experiment

In order to show that a tetrahedral intermediate was involved, Myron Bender (in 1951) investigated the hydrolysis of ethyl benzoate labelled with the isotope ^{18}O at the carbonyl oxygen. The sodium hydroxide added was unlabelled. If the tetrahedral intermediate really did form, it would seem reasonable that proton transfer swapping the proton between the labelled and unlabelled oxygens could occur (the two structures would have exactly the same energy).

Bender realized that if this intermediate were then to break down to reform the starting ester, that ester molecule would now be unlabelled. So he went looking for a reduction in the extent of labelling in the *unreacted* ester. Finding exactly that provided strong evidence for the tetrahedral intermediate.

AMINES AND AMINO ACIDS

24.1 INTRODUCTION TO AMINES AND AMINO ACIDS

- Amines are one class of organic compounds that contain **nitrogen** atoms.
- We classify amines as shown in Figure 24.1:
  - **Primary amines**, with the general formula RNH_2, have one alkyl group and two hydrogens attached to the nitrogen atom.
  - **Secondary amines**, with the general formula RR'NH, have two alkyl groups and one hydrogen attached to the nitrogen atom.
  - **Tertiary amines**, with the general formula RR'R"N, have three alkyl groups attached to the nitrogen atom.
  - **Quaternary ammonium salts** contain the ion $RR'R''R'''N^+$ which has four alkyl groups attached to the positively charged nitrogen.

Note carefully that the use of the terms 'secondary' etc. **differs** from their use in both halogenoalkanes (Section 20.2) and alcohols (Section 21.4). It is the number of alkyl groups bonded to the **nitrogen**, rather than the **carbon** bonded to that nitrogen, that matters. For example, $(CH_3)_3COH$ is a *tertiary* alcohol, whereas $(CH_3)_3CNH_2$, as shown in Figure 24.2, is a *primary* amine.

- There are two common naming systems for amines:
  - In the first, we name the alkyl group(s) and add the ending '-amine'.
  - In the second, we treat the $-NH_2$ group (**amino group**) as a substituent.
    - So we can call $CH_3CH_2NH_2$ ethylamine or aminoethane.
    - We can call the molecule in Figure 24.2 2-amino-2-methylpropane.

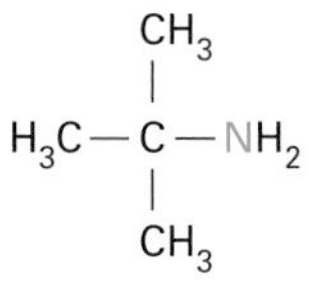

24.2 AMINES AS BASES

- Because of the lone pair on nitrogen, amines are **Lewis bases** (Section 7.6). Lewis bases are always **Brønsted bases** (Section 7.1). So, in a similar way to ammonia, the lone pair can accept a proton from a Brønsted acid.
- **Primary amines are slightly more basic than ammonia** because the electron-donating alkyl group (Section 18.3) stabilizes the positive ion formed:

Other examples of organic nitrogen-containing compounds include, for example, nitriles RCN, urea $CO(NH_2)_2$, and proteins.

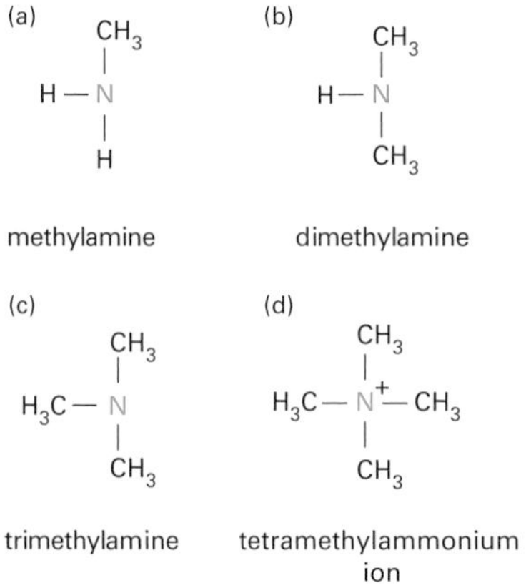

FIGURE 24.1 Amines: (a) the primary amine methylamine, (b) the secondary amine dimethylamine, (c) the tertiary amine trimethylamine. (d) is the tetramethylammonium ion.

FIGURE 24.2 The primary amine *t*-butylamine: Section 16.2 explains the notation '*t*-butyl'.

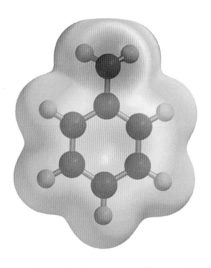

FIGURE 24.3 The electrostatic potential map for phenylamine (aniline). The red coloration, showing high electron density, identifies the nitrogen lone pair. The yellow colour, showing relatively high electron density, spreads out towards the benzene ring.

$$- \quad RNH_2 + H_2O \rightleftharpoons RNH_3^+ + OH^-$$

- **Aromatic amines are generally *much less basic*** than the corresponding aliphatic amines. The nitrogen lone pair is delocalized with the benzene ring, as it aligns with the p orbitals of the carbon atoms in the ring. This can be seen from the electrostatic potential map shown in Figure 24.3. Protonation destroys the delocalization as there is then no lone pair. **Phenylamine (aniline)**, $C_6H_5NH_2$, is a weaker base than methylamine, CH_3NH_2, by 6 orders of magnitude.

- We can quantify the basicity using the pK_a of the conjugate acid; see Taking the Next Step 24.1.

TAKING THE NEXT STEP 24.1

Using pK_a values for the conjugate acid

We can quantify the basicity of an amine using the measured pK_a value (Section 7.3) of its *conjugate acid* (Section 7.1). Take, for example, the conjugate acids of ammonia, methylamine, and phenylamine:

$$NH_4^+ + H_2O \rightleftharpoons H_3O^+ + NH_3 \qquad pK_a = 9.3$$

$$CH_3NH_3^+ + H_2O \rightleftharpoons H_3O^+ + CH_3NH_2 \qquad pK_a = 10.6$$

$$C_6H_5NH_3^+ + H_2O \rightleftharpoons H_3O^+ + C_6H_5NH_2 \qquad pK_a = 4.6$$

The *lower* the pK_a value, the *more* acidic an acid is (that is, the more likely it is to donate a proton, Section 7.1) and the further the equilibrium lies towards products, which in this case corresponds to the amine itself. The difference in basicity between phenylamine and methylamine corresponds to a difference of 6.0 in the pK_a values of the conjugate acid, which being logarithms (Section 7.3) equates to 6 orders of magnitude. The electron-donating methyl group (Section 18.3) stabilizes the conjugate acid of methylamine compared with ammonia by a little over 1 order of magnitude.

> From what you came across pre-university, you may not have appreciated the point that the delocalization effect that stabilizes the aromatic amine is often ***far*** more significant than the electron-donating effect (often called the **inductive** effect).

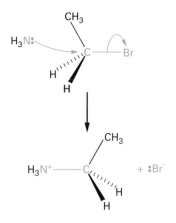

FIGURE 24.4 The mechanism of the reaction between ammonia and bromoethane. Deprotonation follows to form the amine itself.

24.3 AMINES AS NUCLEOPHILES AND THEIR PREPARATION

- We can make **primary amines** by the nucleophilic substitution reaction (Section 20.2), shown in Figure 24.4, between ammonia (NH_3) and a halogenoalkane (RX).

- However, because of the lone pair on nitrogen, we can also expect amines to be **nucleophiles**, in a similar way to ammonia. Hence when ammonia (NH_3) reacts with a halogenoalkane (RX) to form a primary amine (RNH_2), the reaction can continue to form the secondary amine (R_2NH) by attack of the nucleophilic primary amine, and so on.

- So this reaction is poor for making primary amines since it is too difficult to prevent further alkylation. A Deeper Look 24.1 discusses one way of making primary amines in good yield.

- Amines also react with acyl chlorides in a very similar way to ammonia (Section 23.3).

A DEEPER LOOK 24.1

What is the Gabriel synthesis?

The **Gabriel synthesis** is a way of making a *primary* amine much more successfully. The nitrogen atom in the **Gabriel reagent** (potassium phthalimide) carries a full negative charge and is therefore a more powerful nucleophile than the nitrogen atom in ammonia. Most importantly, once the first alkyl group has attached, as shown in Figure 24.5, no further deprotonation can occur; the lone pair on nitrogen is delocalized with the two neighbouring carbonyl groups (Section 2.3) and so is not available for further reaction.

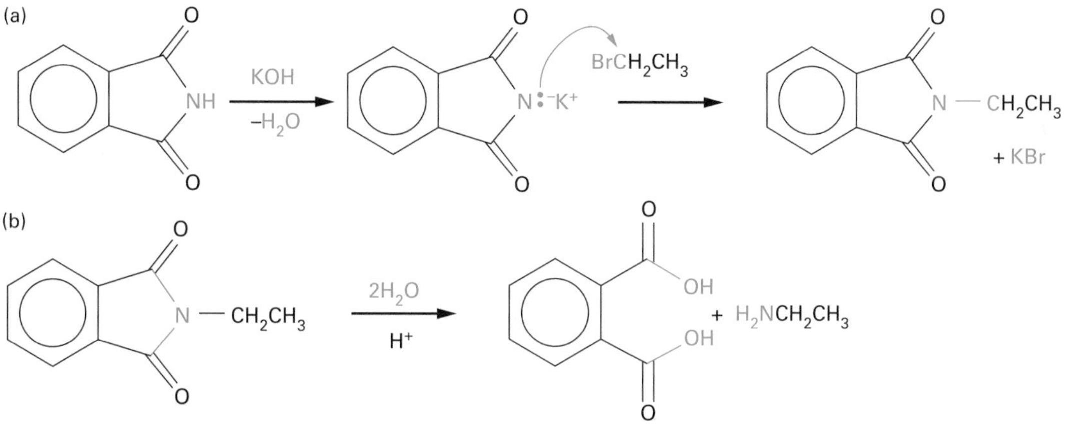

FIGURE 24.5 The Gabriel synthesis of an amine.

A DEEPER LOOK 24.2

Why do halogenoarenes not react as halogenoalkanes do?

To perform an S_N2 reaction (Section 20.2), the nucleophile would have to approach directly *through* the aromatic ring, which is impossible geometrically. (Negatively charged nucleophiles would also be repelled electrostatically by the high electron density in a benzene ring.) As a result, chlorobenzene does not react with ammonia to give phenylamine; nor does chlorobenzene react with hydroxide ion to give phenol.

24

To perform an S_N1 reaction (Section 20.2), an aryl carbocation would need to form, which is impossible unless the exceptionally good leaving group N_2 is present (see A Deeper Look 24.3).

At university, you will learn that *nucleophilic* aromatic substitution is in fact possible, for example by an addition-elimination mechanism, but even then only if the aromatic ring has an electron-withdrawing group (such as $-NO_2$) *ortho* or *para* (Figure 19.18) to the halogen.

A DEEPER LOOK 24.3

What is diazotization?

The **diazotization reaction** uses nitrous acid HNO_2 (HONO) to convert phenylamine $C_6H_5NH_2$ into the diazonium salt **benzenediazonium chloride** $C_6H_5N_2^+Cl^-$. Nitrous acid decomposes very readily and so cannot be stored. It is prepared *in situ* (within the reaction mixture) from a mixture of sodium nitrite $NaNO_2$, and an acid, i.e.

$$NaNO_2(aq) + HCl(aq) \rightarrow HNO_2(aq) + NaCl(aq)$$

The temperature must be kept below 5 °C in an ice-bath during diazotization, otherwise the diazonium ion decomposes, losing its nitrogen atoms as nitrogen gas N_2. Temperature control is critical, because below 0 °C the rate of diazotization becomes very slow.

The benzenediazonium ion, as shown in Figure 24.6, reacts with nucleophiles. For example, water will attack the aryl carbocation formed after loss of nitrogen gas, as shown in Figure 24.7. (The production of nitrogen *gas* with its very strong triple bond makes N_2 an exceptional leaving group.) The result of the reaction (after deprotonation) is **phenol** (hydroxybenzene) C_6H_5OH: this is in fact the best way to make phenol in the laboratory. Iodide ion (from potassium iodide) will also react to form **iodobenzene C_6H_5I**.

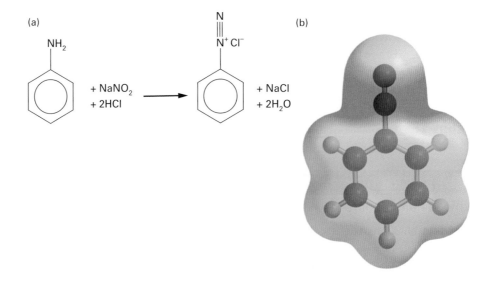

FIGURE 24.6 (a) Formation of benzenediazonium chloride; the nitrogen attached to the ring carries the major part of the positive charge, (b) as shown by the dark blue colour in the electrostatic potential map.

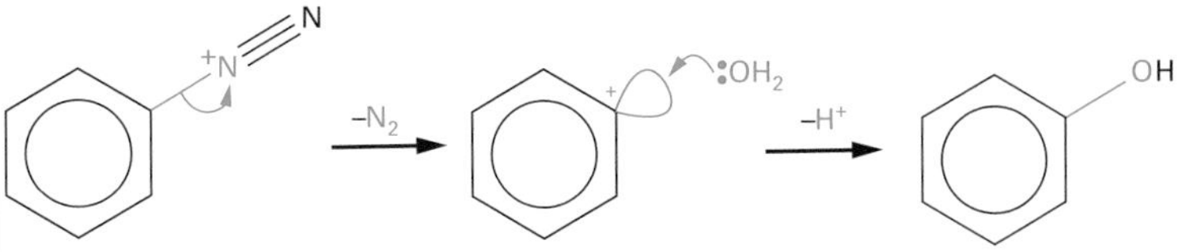

FIGURE 24.7 The formation of nitrogen gas with its very strong triple bond makes N_2 an exceptional leaving group. Water then attacks the aryl carbocation, which forms phenol after deprotonation.

- **Phenylamine** is *not* made in a similar way (A Deeper Look 24.2 explains why). Instead we can make it by reducing nitrobenzene: in the reduction of $C_6H_5NO_2$ to $C_6H_5NH_2$, oxygen is lost and hydrogen gained.
  - A specific reducing agent commonly used is **tin** and **concentrated hydrochloric acid**. (The metal tin provides the electrons used for reduction.)

24.4 AMINO ACIDS

- **Amino acids** contain two functional groups: an amino group ($-NH_2$) and a carboxylic acid group ($-COOH$).
  - The most common amino acids are the **alpha(α)-amino acids** where the two groups are attached to the *same* carbon atom.
- There are **20 naturally occurring** α-amino acids $RCH(NH_2)COOH$.

    During university chemistry courses, proline may be more accurately described as an imino acid as it has a secondary amino group, as seen in Figure 24.8.

  - Table 24.1 shows some examples. All of these examples, apart from glycine, have at least one asymmetric carbon atom: see Section 16.3

An important amino acid that is not an α-amino acid is the neurotransmitter **GABA** (gamma-aminobutanoic acid) $NH_2CH_2CH_2CH_2COOH$.

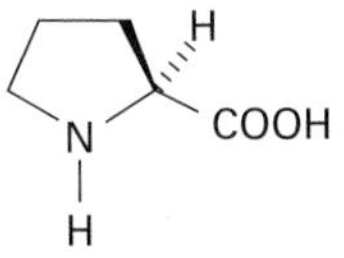

FIGURE 24.8 Proline is an imino acid.

TABLE 24.1 Some of the amino acids occurring in proteins.

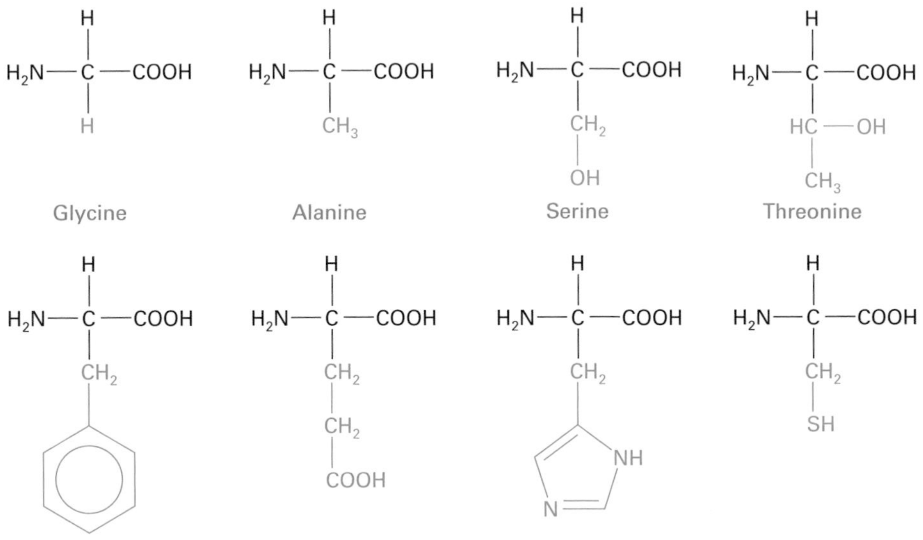

| Glycine | Alanine | Serine | Threonine |

| Phenylalanine | Glutamic acid | Histidine | Cysteine |

24

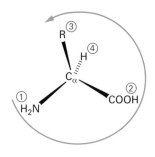

FIGURE 24.9 Almost all the natural amino acids have the *S*-configuration (Section 16.3). However, in cysteine the R group (CH_2SH) takes priority over COOH, as S has a higher atomic number than O.

for an explanation of chirality. All naturally occurring amino acids (apart from cysteine) have the *S*-configuration, as shown in Figure 24.9.

24.5 EFFECT OF pH AND ZWITTERIONS

- Because of the basic amino group, **at pH 0** the amino acid is **protonated** to form the ion $RCH(NH_3^+)COOH$. (The carboxylic acid group remains protonated.)

- Because of the acidic carboxylic acid group, **at pH 14** the amino acid is **deprotonated** to form the ion $RCH(NH_2)CO_2^-$. (The amino group remains unprotonated.)

- Near neutral pH, the carboxylic acid group in the molecule is deprotonated while its amino group is protonated to form a **zwitterion** $RCH(NH_3^+)CO_2^-$, as shown in Figure 24.10.

- Zwitterion formation results in a generally high melting point for the amino acids.

24.6 AMINO ACIDS POLYMERIZE TO FORM POLYPEPTIDES

- The bond linking two amino acids together is a substituted amide (Section 23.3) bond, commonly called a **peptide bond** (more logically called a **peptide link**, as five actual covalent bonds are involved), as shown in Figure 24.11.

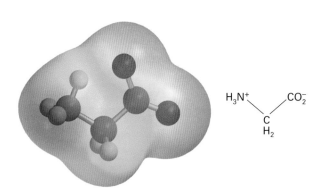

$$H_3N^+ \quad CO_2^-$$
$$\underset{H_2}{C}$$

FIGURE 24.10 Most amino acids exist as zwitterions at physiological pH: the protonated amine group is blue and the carboxylate group is red.

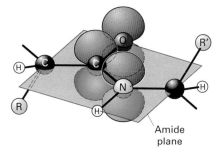

Amide plane

FIGURE 24.11 The p orbitals on carbon and oxygen that form the π bond can also overlap with the p orbital on nitrogen, when this is arranged as in the figure to cause a structure delocalized over all three atoms. This delocalization is the reason why the peptide bond is planar.

- Because of delocalization (Section 2.3), as seen in Figure 24.11, the **peptide bond is planar**. The overlap and consequent delocalization is a maximum when the p orbital on the nitrogen atom is oriented perpendicular to the plane.

- Amino acids can bond together to form **condensation polymers** (Section 25.1): we call the polymers **polypeptides**. **Proteins** consist of one or more polypeptide chains: see Taking the Next Step 24.2.

Oligopeptides are smaller molecules with up to twenty amino acids bonded. **Dipeptides** and **tripeptides** have two and three amino acids bonded respectively.

TAKING THE NEXT STEP 24.2

What is the structure of a polypeptide?

We can describe polypeptide structure at four levels.

The **primary structure** of a polypeptide describes the sequence of amino acids present: peptide bonds join the amino acids together. It would be tedious to write out the full structure for each amino acid, so we can use a three-letter code: Gly stands for glycine, Ala for alanine etc. Figure 24.12 shows the primary structure of myoglobin. A one-letter code is used *much* more frequently now (however, for example, K for lysine is much less obvious than Lys).

$$\underset{\text{1}}{H_2N} - \underset{}{Val} - Leu - Ser - Glu - \underset{5}{Gly} - Glu - Trp - Gln - Leu - \underset{10}{Val} - Leu -$$

$$Ala - \underset{145}{Ala} - Lys - Tyr - Lys - Glu - Leu - \underset{150}{Gly} - Tyr - Gln - Gly - COOH$$

FIGURE 24.12 This sequence of amino acids gives part of the primary structure of sperm whale myoglobin.

We can find the sequence of amino acids experimentally using mass spectrometry.

The **secondary structure** of a polypeptide explains how it can fold into characteristic three-dimensional arrangements, stabilized by hydrogen bonding (Section 4.2). Figure 24.13 shows the two most common structures: (a) the **alpha helix** and (b) the **beta pleated sheet**.

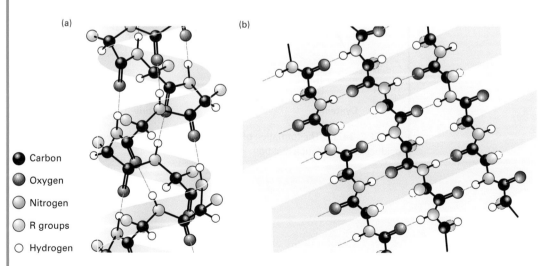

(a) (b)

- Carbon
- Oxygen
- Nitrogen
- R groups
- Hydrogen

24

FIGURE 24.13 Hydrogen bonding, shown here by orange dashed lines, gives rise to the two forms of secondary structure: (a) the alpha helix; and (b) the beta pleated sheet (here, three-stranded).

Proline (the imino acid shown in Figure 24.8), when in a peptide bond, lacks an N-H group and so breaks up an alpha-helical section of the protein. (We call it a helix disrupter.) The proline side chain also interferes with the helical packing.

The **tertiary structure** describes the overall shape of the whole polypeptide chain. Interactions between distant amino acids include disulfide bridges. **Disulfide bridges** are covalent bonds, as shown in Figure 24.14, that form between sulfur atoms upon the oxidation of two cysteines (see Table 24.1).

Figure 24.15 shows the tertiary structure of myoglobin, about 70% of which is folded into α helices. Amino acids with neutral, non-polar side chains have a strong tendency to accumulate in the interior of the molecule: we describe these amino acids as **hydrophobic** ('water-hating'). By contrast, those that have side chains that can become charged tend to congregate on the outside where they can interact with water: we describe these amino acids as **hydrophilic** ('water-loving').

Finally, some proteins have a **quaternary structure**, which occurs only if there is more than one polypeptide subunit involved and refers to their spatial arrangement. Haemoglobin, for example, consists of two pairs of identical polypeptide subunits (see Figure 15.21).

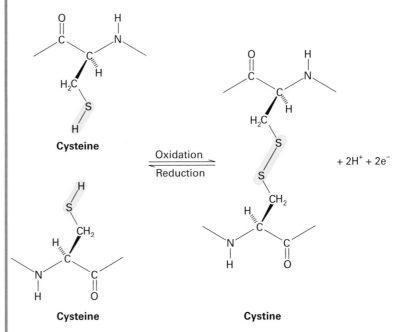

FIGURE 24.14 The formation of a disulfide bridge between two cysteine residues is an oxidation reaction.

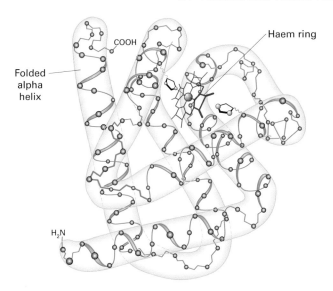

FIGURE 24.15 The tertiary structure of myoglobin. The interior consists almost entirely of non-polar amino acids (apart from the two histidines shown in cartoon form): the polar ones are on the outside.

- We can **hydrolyse** proteins into their constituent amino acids using strong acid (or enzymes).

- Almost all **enzymes** are proteins.

> **Ribozymes** are ribonucleic acid (RNA) enzymes that have catalytic properties; they were first identified in *E. coli* in 1981.

24

CHAPTER
25

POLYMERS

25.1 INTRODUCTION TO THE POLYMERS

- A **polymer** is a very large molecule formed by reacting together *many* small molecules called **monomers**.

- We can classify polymers as addition polymers or condensation polymers.

- We can make **addition polymers** by adding together many small molecules: no atoms or molecules are produced as by-products. By far the most important examples are the **polyalkenes** (one example is shown in Figure 25.1; see Section 25.2).

- The **repeating unit** of a polymer specifies the smallest part of the molecule which on replication multiple times reproduces the structure of the polymer.

- **Condensation polymers** are made by reacting together two difunctional compounds. **Difunctional** compounds have two functional groups (in this instance, often two identical groups). A small molecule, such as water, will also be produced when reaction occurs.

  - Condensation polymers are usually either polyesters, as shown in Figure 25.2, or polyamides.

At university, you may come across the term 'bifunctional' used when the two functional groups are different, with the term 'difunctional' reserved for the case where the two functional groups are the same.

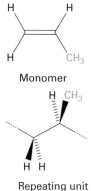

FIGURE 25.1
The monomer and repeating unit of poly(propene).

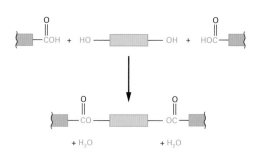

FIGURE 25.2 A polyester forms from the reaction between two different monomer molecules.

25.2 ADDITION POLYMERS: POLYALKENES

- **Addition polymerization** of alkenes involves the formation of one huge molecule by the joining together of a large number of alkene monomers.

- The monomer of **poly(ethene)** is ethene.

  - Poly(ethene) is also frequently called **polyethylene** or **polythene**.

  - Two different forms of poly(ethene) exist. One form (**LDPE, low-density polyethylene**) is made at very high pressures: see A Deeper Look 25.1. The other form (**HDPE, high-density polyethylene**) is made using a catalyst mixture discovered by Karl Ziegler and Giulio Natta.

    The **Ziegler–Natta catalyst** is typically a mixture of titanium(III) chloride $TiCl_3$ and triethylaluminium $Al(CH_2CH_3)_3$. See A Deeper Look 25.2.

- Figure 25.3 shows four other commercially important addition polymers—poly(propene) or polypropylene (**PP**), poly(chloroethene) or polyvinyl chloride (**PVC**), poly(phenylethene) or polystyrene (**PS**), and poly(tetrafluoroethene) or **PTFE**—together with the repeating unit in each case. Taking the Next Step 25.1 discusses stereoregular polymers.

A sample of a polymer such as poly(ethene) consists of a mixture of molecules with different molar masses.

Le Chatelier's principle (Section 6.3) explains why high pressure favours the formation of a solid from gaseous reactants.

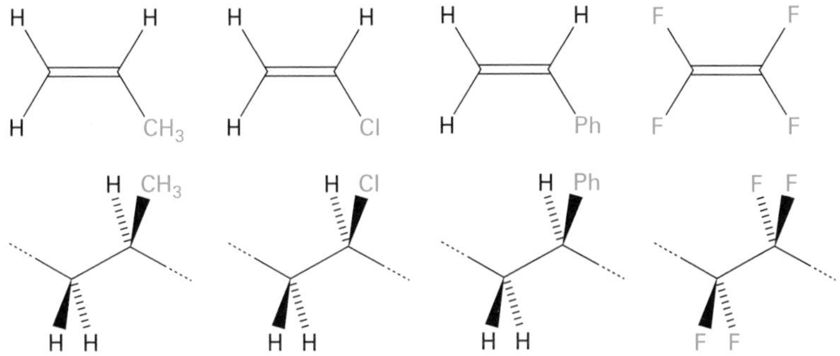

FIGURE 25.3 The monomers are propene (propylene), chloroethene (vinyl chloride), phenylethene (styrene), and tetrafluoroethene. The traditional names for the polymers are polypropylene (PP), polyvinyl chloride (PVC), polystyrene (PS), and PTFE.

The **IUPAC name** for the polymer is very easy: put the name of the alkene which is its monomer in brackets after the prefix 'poly'.

TAKING THE NEXT STEP 25.1

What are stereoregular polymers?

Polymerization of propene (for example) with a Ziegler–Natta catalyst can result in an **isotactic polymer**, in which all the methyl groups are on the same side of the chain. Figure 25.4(a) shows such a polymer. We can also modify the catalyst mixture to produce a **syndiotactic polymer**, in which the methyl groups alternate regularly from one side of the chain to the other. Figure 25.4(b) shows such a polymer. We call isotactic and syndiotactic polymers **stereoregular polymers**.

An **atactic polymer** has the methyl groups distributed randomly: the high-pressure synthesis forms an atactic polymer. Compared with the atactic polymer, the isotactic polymer is harder, more crystalline, has a higher melting point, and has a lower solubility in most solvents. Jug kettles are made from isotactic polypropylene.

25

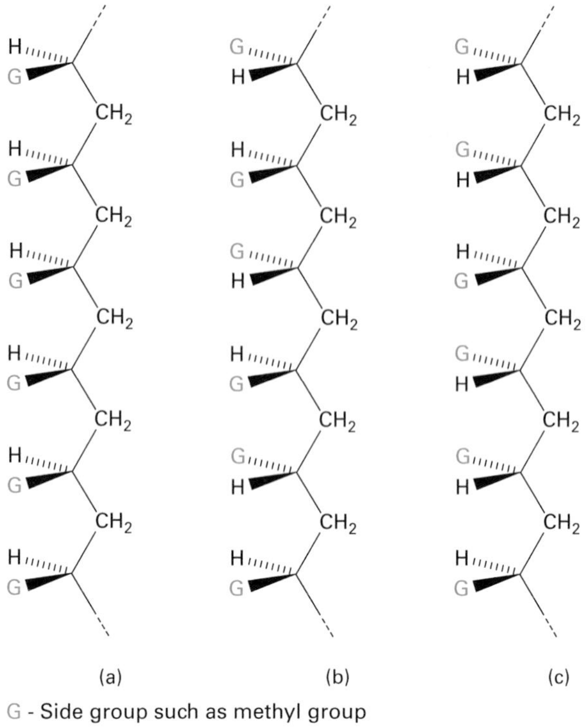

(a)   (b)   (c)

G - Side group such as methyl group

FIGURE 25.4 Polymer geometry: (a) isotactic (all Gs on the same side of the hydrocarbon chain); (b) syndiotactic (alternating sides); and (c) atactic (random sides).

A DEEPER LOOK 25.1

What is the mechanism for formation of LDPE?

The high-pressure commercial production of poly(ethene) involves radicals (Section 16.4). Figure 25.5 shows the chain initiation, propagation, and termination steps.

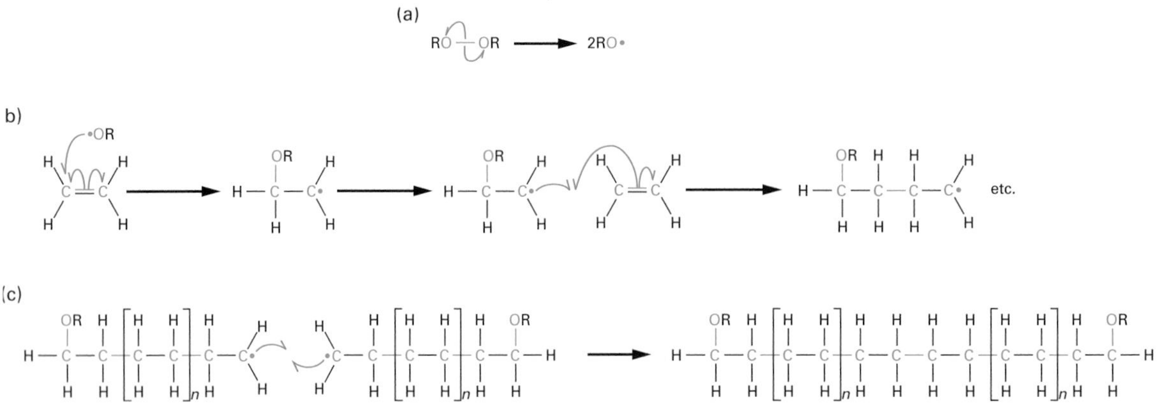

FIGURE 25.5 The mechanism of commercial high-pressure addition polymerization (1400 atm and 170 °C) to form LDPE. (a) Initiation: radicals are generated from the decomposition of an organic peroxide. (b) Propagation: ethene adds to the radical; the unpaired electron is situated on the end carbon atom. (c) Termination: two radicals combine.

25

A DEEPER LOOK 25.2

What is the mechanism for formation of HDPE?

The mechanism for production of HDPE using Ziegler–Natta catalysts is complex. Figure 25.6 shows a *simplified* version. The ethyl group is transferred from the aluminium triethyl co-catalyst. HDPE has a very small degree of chain branching. As a result, the polymer chains pack more closely than in LDPE, which has more branched chains. HDPE is therefore harder and more rigid than LDPE. See also Taking the Next Step 25.1.

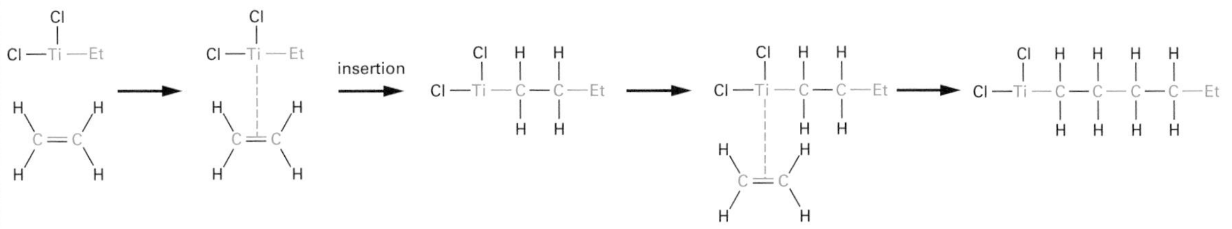

FIGURE 25.6 A summary of the Ziegler–Natta polymerization of ethene to form HDPE. We call the mechanism **coordination polymerization**, because coordination compounds (complexes) form as electron density from the ethene π bond is donated into d orbitals on an atom of titanium (the catalyst). One end of the growing hydrocarbon chain is attached to a titanium atom, while incoming ethene molecules form a complex with the same titanium atom. We call the second step an **insertion** reaction.

25.3 CONDENSATION POLYMERS: POLYESTERS

- We usually make **polyesters** by reacting together a dicarboxylic acid (or a diacyl chloride or a diester) and a diol, as shown in Figure 25.7. The links between the units in the polymer are ester links (Section 23.4).

A single difunctional molecule bearing both a carboxylic acid group and a hydroxyl group can also act as a monomer.

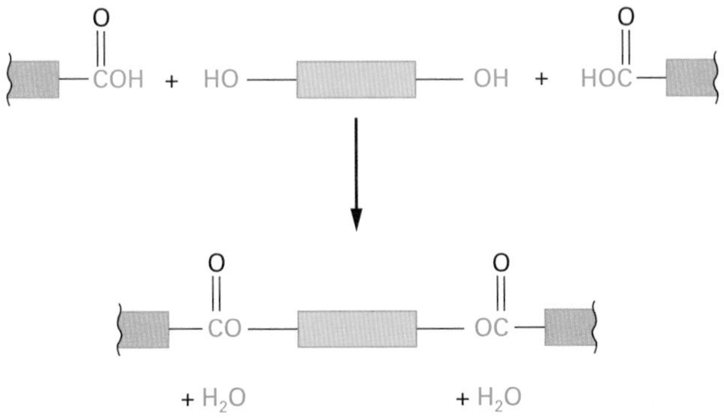

FIGURE 25.7 A polyester forms from the reaction between two different monomer molecules.

Figure 25.8 shows the very important polyester **PET** (polyethylene terephthalate) or **Terylene** (Dacron in the US), used for permanent-press fabrics for example.

- Terylene's repeating unit is $-OCH_2CH_2OCOC_6H_4CO-$. One ester link is highlighted in red.

25

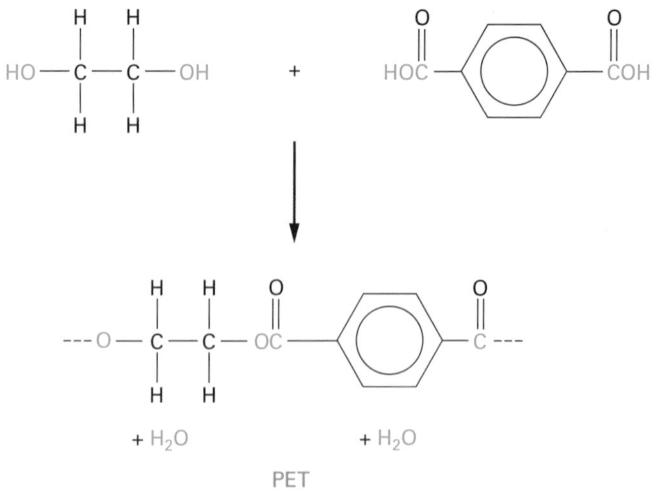

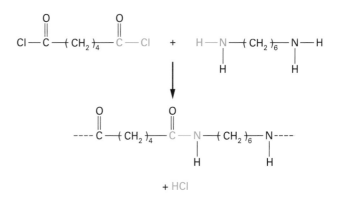

FIGURE 25.8 The industrial production of PET often uses the dimethyl ester instead of the free acid.

25.4 CONDENSATION POLYMERS: POLYAMIDES

Other natural polymers include **polysaccharides** and **nucleic acids**.

We usually make **polyamides** by reacting together a dicarboxylic acid (or a diacyl chloride) and a diamine. The links between the units in the polymer are substituted amide links (Section 23.3). **Proteins** (Section 24.6) are natural polyamides.

Figures 25.9 and 25.10 show the most important polyamide **nylon 6,6**, used for textiles and carpets for example.

The synthesis of nylon is interesting as it occurs at the interface of two immiscible liquids. The diacid only dissolves in water; the diamine only dissolves in organic solvents. Where the solutions touch, i.e. the interface, reaction can occur. Many textbooks refer to this as the **nylon rope trick**, as illustrated in Figure 25.9, as you can grab the end of the fibre and keep pulling it out, since more chemicals come into contact as the reacted ones are removed. We name nylons by describing the number of carbons in each monomer, e.g. nylon 6,6 uses the hexyl diacid and the hexyl diamine.

FIGURE 25.9 The nylon rope trick.

FIGURE 25.10 The condensation reaction between hexane-1,6-dioyl chloride and 1,6-diaminohexane forms nylon 6,6.

Figure 25.10 shows how we can form nylon 6,6. The repeating unit of nylon 6,6 is

$-OC(CH_2)_4CONH(CH_2)_6NH-$. One amide link is highlighted in red.

Taking the Next Step 25.2 covers aramid fibres. A Deeper Look 25.3 explains one final distinction, between chain and step polymerization.

TAKING THE NEXT STEP 25.2

Aramid fibres

Aramid fibres are a class of polyamides in which the amide links are directly attached to benzene rings. Attaching the amide links in the 1 and 4 positions gives a very strong material called **Kevlar**, which is used to make bullet-proof vests and (combined with carbon fibre as CarboKevlar) some parts of F1 racing cars such as the petrol tank. (The linear arrangement at the benzene ring contributes to its tensile strength. When spun into fibres the chains orientate themselves along the fibre axis by hydrogen bonding between adjacent chains.) We can make Kevlar by reacting benzene-1,4-dicarboxylic acid with benzene-1,4-diamine. Figure 25.11 shows the repeating unit of Kevlar.

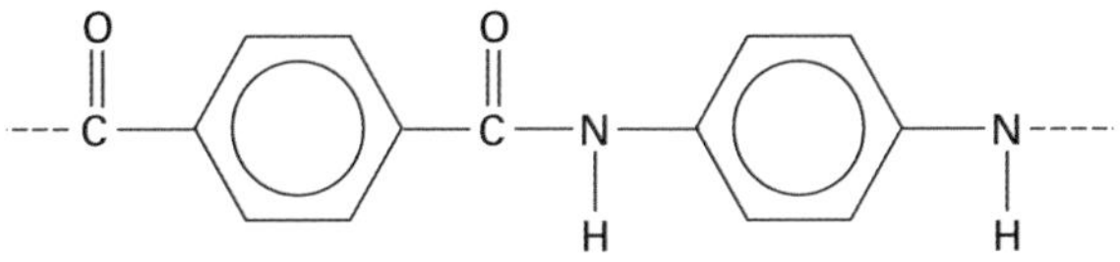

FIGURE 25.11 The repeating unit of Kevlar.

A DEEPER LOOK 25.3

Chain and step polymerization

Poly(ethene), along with other addition polymers, is formed by **chain polymerization**: monomers add to the reactive site of a growing chain followed by regeneration of the reactive site, as shown in Figure 25.5. During chain polymerization, the monomer concentration falls steadily over time, forming very long polymer chains rapidly. The yield of polymer increases with reaction time but the polymer molar mass does not change significantly.

In contrast, condensation polymers such as polyesters and polyamides are formed by **step polymerization**: any two monomers or short chains can react. The monomer concentration drops very rapidly and the polymer chains are initially relatively short. Long polymer chains require long reaction times. The reaction mixture contains a collection of polymers with variable molar mass. **Polyurethanes** (PUR or PU) are also made by step polymerization but they are not condensation polymers (as no small molecule forms as a by-product), as seen in Figure 25.12.

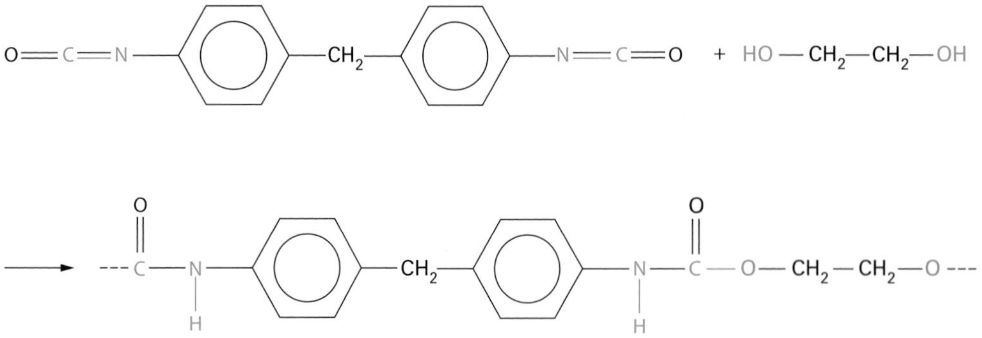

FIGURE 25.12 The production of polyurethane (the **urethane** linkage is −NHCOO−).

INSTRUMENTAL ANALYSIS

- **Instrumental techniques** used for analysis in chemical laboratories, especially those working with organic compounds, include mass spectrometry (MS), infrared (IR) spectroscopy, and nuclear magnetic resonance (NMR) spectroscopy.

  > A technique described as **spectroscopy** must involve interaction between the sample and **electromagnetic radiation**. As the typical mass spectrometry technique does not involve electromagnetic radiation, it is *not* a spectroscopic technique.

26.1 MASS SPECTROMETRY

- We can use mass spectrometry (Section 1.2) to find a compound's **molecular formula** because each individual element has a mass that is known precisely to at least four decimal places. Hence measuring the mass of the **molecular ion** (the ion formed by loss of one electron from the whole molecule) to four decimal places (**high resolution mass spectrometry**) can distinguish between molecules that have the same mass to the nearest integer. The **base peak** is the most intense peak, as shown in Figure 26.1.

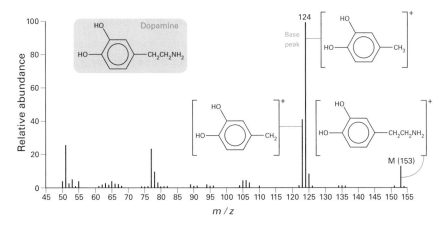

FIGURE 26.1 A (simplified) mass spectrum of dopamine, which is a neurotransmitter in the brain. The molecular ion peak at $m/z = 153$ corresponds to an ion of formula $C_8H_{11}O_2N^+$. The peak at $m/z = 124$ (the base peak) corresponds to the ion $C_7H_8O_2^+$. This ion results from fission of the $C_6H_3(OH)_2CH_2{-}CH_2NH_2$ bond (producing $C_6H_3(OH)_2CH_2^+$) followed by capture of a hydrogen atom from other molecular fragments. This structure allows delocalization of the positive charge with the benzene ring π cloud.

If a molecule contains an odd number of nitrogen atoms, its molecular ion will have an odd m/z value.

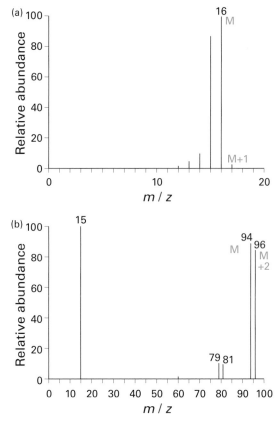

FIGURE 26.2 (a) The mass spectrum of methane CH_4 showing the molecular ion peak at $m/z = 16$ and the (much smaller) M+1 peak at $m/z = 17$. (b) The (simplified) mass spectrum of bromomethane showing the M+2 peak.

At university, you will also have to consider the possibility of having deuterium (D or ^2H) present.

- Figure 26.2(a) shows that the ratio of the **(M+1) peak** to the M peak (M being the molecular ion) determines the number of carbon atoms. Because the isotope carbon-13 is present at 1% abundance, an 8% ratio (in octane for example) means there are 8 carbon atoms.

- We can spot the presence of a halogen atom quickly because the ratio of the M peak to the **(M+2) peak** will be 3:1 if one chlorine atom is present or 1:1 if one bromine atom is present, because the isotopic ratio of ^{35}Cl to ^{37}Cl is close to 3:1, while that for ^{79}Br to ^{81}Br is close to 1:1, as shown in Figure 26.2(b).

- Taking the Next Step 26.1 discusses fragmentation.

TAKING THE NEXT STEP 26.1

What is fragmentation?

The molecular ion can fragment (break apart) as it flies through the spectrometer. The **fragmentation pattern** is characteristic of the molecule; a match to stored mass spectra of known compounds on a computer database enables rapid identification of the sample.

The most common fragment peaks seen in the mass spectrum are those that form stable positive ions. We can therefore expect **tertiary carbocations**

(Section 18.3) to be seen: $(CH_3)_3C^+$ has a mass of 57. There is a peak at m/z = 57 in Figure 26.3(b) (although it is not the base peak). Another common type of fragment is the **acylium ion** RCO^+ (Section 19.4), often seen in the mass spectrum of a carbonyl compound, as shown in Figure 26.4.

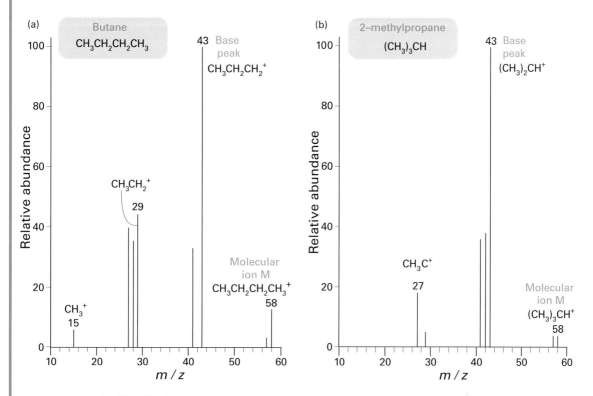

FIGURE 26.3 The (simplified) mass spectra of two isomers. (a) Butane. The main fragments are identified. (b) 2-Methylpropane. Note the following two major changes. The peak at m/z = 29 is very much smaller, as the fragment $CH_3CH_2^+$ cannot be formed. The relative heights of the peaks at m/z = 57 and 58 are the same, because breaking the $(CH_3)_3C–H$ bond creates a tertiary carbocation, which is relatively stable. As for butane, the base peak is formed by the loss of one methyl group.

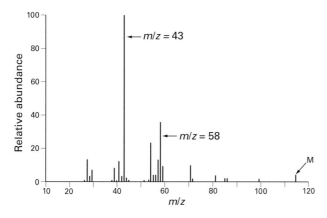

FIGURE 26.4 The mass spectrum of 5-methylhexan-2-one. The base peak at m/z = 43 is due to CH_3CO^+. Note that the molecular ion peak is small, as is also the case for both molecules shown in Figure 26.3.

26.2 INFRARED SPECTROSCOPY

- **Infrared (IR) spectroscopy** involves the absorption of radiation in the infrared region of the electromagnetic spectrum; the energy absorbed makes **bonds vibrate** (stretch or bend) more energetically.
  - Hence **IR identifies bonds**.
- We report the absorption (conventionally recorded as the **transmittance** through the sample) in an unusual unit called **wavenumber** (measured in cm^{-1}). The typical range is from **4000 to 400 cm^{-1}** (2000 cm^{-1} corresponds to a wavelength of 5 µm). Higher wavenumbers correspond to higher vibration frequencies.
- Table 26.1 shows the typical range of wavenumbers where absorption due to stretching vibrations appears.
- The **largest wavenumber values (3600 to 2500 cm^{-1}) involve bonds to hydrogen** as hydrogen atoms are the lightest atoms and so vibrate fastest.
  - Absorption due to an **O−H bond** is usually broad (due to hydrogen bonding).
- The **next largest values are associated with double bonds**, as the extra pair of electrons pulls the atoms back together more forcibly.
  - **Compounds that have C=O bonds generally absorb at between 1680 and 1750 cm^{-1}.** See Section 22.2 for the difference between simple

We can distinguish an intramolecular hydrogen bond from an intermolecular one because the former does not depend on the concentration of the sample.

TABLE 26.1 Typical infrared wavenumbers.

Bond	Type of compound	Range/cm^{-1}	Bond	Type of compound	Range/cm^{-1}
C−H	Alkanes	2850−2960	C=C	Alkenes	1620−1680
	Alkenes	3010−3095	C=C	Arenes	1500−1600
	Arenes	3030−3080	C−O	Alcohols, ethers, carboxylic acids, esters	1000−1300
	Aldehydes	2700−2770	C=O	Aldehydes, ketones, carboxylic acids, esters	1680−1750
O−H	Alcohols (H-bonded)	3230−3550			
	Carboxylic acids (H-bonded)	2500−3000	C−N	Amines	1180−1360
N−H	Amines	3320−3560	C≡N	Nitriles	2210−2260

You need to be aware that this list suggests where a particular stretch will occur rather than being a definitive list. The actual wavenumber seen will depend on the exact details of the molecular structure.

aldehydes and simple ketones. A Deeper Look 23.1 describes how to use the value to assess the extent of delocalization of the carbonyl group with the lone pair on attached atoms.

- Absorptions below 1500 cm^{-1} are usually associated with very complicated motions of the molecule as a whole. Hence this **fingerprint region (1500 to 400 cm^{-1})** is characteristic of a particular molecule, enabling positive identification by comparison with known compounds stored on a computer database.

  > Within the fingerprint region we can still observe some common features. For example, out-of-plane C−H bending occurs at around 700 cm^{-1} in Z-alkenes but at around 970 cm^{-1} in E-alkenes. Similarly, *para*-disubstituted benzenes absorb around 830 cm^{-1}, while *ortho*-disubstituted benzenes absorb around 750 cm^{-1}.

- Bending vibrations require less energy than stretching vibrations, so bending vibrations are typically found within the fingerprint region.

- Figures 26.5 and 26.6 show examples of IR spectra. Notice in particular the prominent peaks for C=O in both figures and for O−H in Figure 26.6.

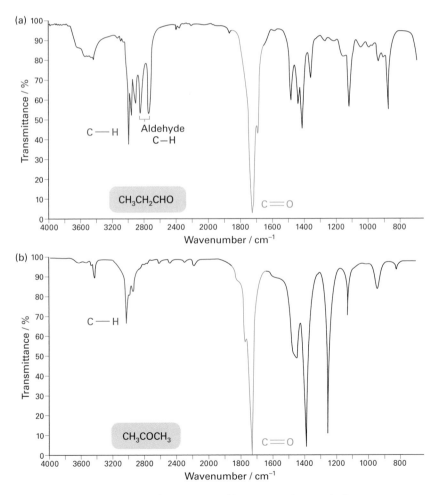

FIGURE 26.5 IR spectra of (a) propanal and (b) propanone (acetone). The most prominent feature in both spectra is the carbonyl C=O stretch. The aldehyde C−H stretch appears at slightly lower frequency than the saturated C−H stretches. The aldehyde C−H stretch often consists, as in (a), of two absorptions (a **doublet**) around 2850 and 2750 cm^{-1}.

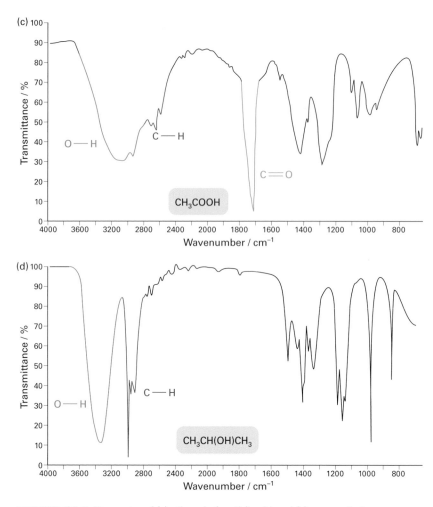

FIGURE 26.6 IR spectra of (a) ethanoic (acetic) acid and (b) propan-2-ol.

26.3 NUCLEAR MAGNETIC RESONANCE (NMR) SPECTROSCOPY

- All nuclei that contain an odd number of protons and/or neutrons (notably those of ^1H and ^{13}C) possess spin: **nuclear spin** is exactly analogous to electron spin (Section 1.4). Spin is an intrinsic angular momentum associated with a particle. Although the picture of a nucleus spinning on its axis is too simplistic, it is nevertheless helpful because a spinning charge (all nuclei being charged) behaves like a small magnet.

- We describe electrons as **spin-up** and **spin-down** shown by up and down arrows (Figure 2.9): we can do the same with these 'magnetic' nuclei. A Deeper Look 26.1 explains why only two states exist.

- The spin-up and spin-down states have the same energy in the absence of a magnetic field. However, once we apply an external **magnetic field**, the two states split in energy, as shown in Figure 26.7, and radiation in the **radio frequency** region of the electromagnetic spectrum can cause a transition between the two energy levels.

- The word '**resonance**' in the name arises because absorption of energy from the electromagnetic field occurs best when the energy of the photon (A Deeper Look 1.3) *exactly matches* the energy gap between the two spin states.

- **Figure 26.8 illustrates an NMR spectrometer.** Features of such instruments include:

  - The solvent used in ^{1}H NMR must be **proton-free**: the most commonly used solvents are deuterated solvents such as $CDCl_3$ (deuterated trichloromethane, chloroform).

  - Figure 26.9 shows the compound **tetramethylsilane (TMS)**, which is used as a standard for measuring chemical shifts: TMS provides the zero value.

    - We use TMS because it gives a single large peak (all protons being equivalent) that is more shielded than almost all other peaks (see A Deeper Look 26.2). It is also non-toxic and volatile.

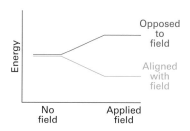

FIGURE 26.7 The spin states of a proton both in the absence and the presence of an applied magnetic field.

FIGURE 26.8 An NMR spectrometer.

A DEEPER LOOK 26.1

Why only two states?

Quantum mechanics explains why an electron (or a ^{13}C or ^{1}H nucleus) with a **spin quantum number** s of ½ can only have one of two possible states. The *magnitude* of the spin (in units of the Planck constant, A Deeper Look 1.3, divided by 2π) is given by the formula $\sqrt{(s(s+1))}$. Its *projection* onto a magnetic field is **quantized** and can have either a value of $m_s = +\frac{1}{2}$ or a value of $m_s = -\frac{1}{2}$. This follows exactly the same rules as for all other forms of angular momentum (compare an electron's magnetic quantum number m_l described in A Deeper Look 1.2).

26.4 ^{13}C (CARBON-13) NMR

- The **number of peaks** indicates the number of different types of carbon atom: **chemically equivalent** carbon atoms (carbon atoms that are in *exactly the same chemical environment*) contribute to the same peak.

- The **chemical shift δ** (measured in parts per million, ppm, relative to TMS) depends on the detailed environment of the particular carbon atom; see Table 26.2.

  - A Deeper Look 26.2 explains the concept of **shielding in NMR.**

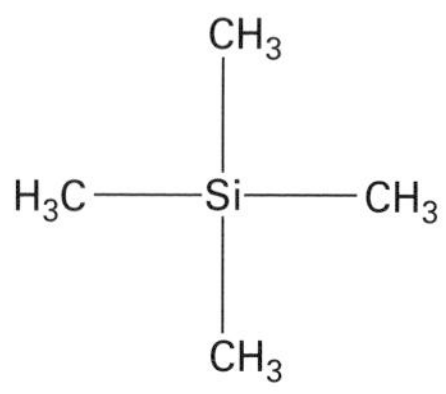

FIGURE 26.9 Tetramethylsilane.

TABLE 26.2 Carbon-13 chemical shifts.

Type of carbon	Chemical shift δ/ppm
Saturated carbons	0–50
Saturated carbons next to O	50–80
Alkene carbons	100–145
Aromatic ring carbons	110–155
C=O (carboxylic acids/derivatives)	155–185
C=O (aldehydes and ketones)	185–220

As was the case for IR stretching frequencies, you should consider this table to be a useful guide rather than a definitive list.

26

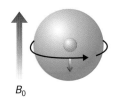

B_0

FIGURE 26.10 A proton surrounded by electron density: the applied field creates an induced magnetic field in the opposite direction.

Although TMS remains the standard for ¹³C NMR, we now normally calibrate the instrument by using the signal from the $CDCl_3$ solvent (at 77 ppm, which we have removed from the spectra below to simplify them).

A DEEPER LOOK 26.2

What is shielding in NMR?

Electron density surrounding a nucleus (for example, a proton) interacts with the applied field B_0. Figure 26.10 shows that the induced magnetic field created is in the opposite direction to B_0.

Electron density (green) surrounds a proton: the external magnetic field causes the electron density to circulate, inducing a small magnetic field in the opposite direction to the external field. This induced field decreases the effect of the external field and '**shields**' the nucleus.

The higher the electron density around the nucleus, the higher the opposing magnetic field and hence the greater the **shielding**. This produces a smaller chemical shift. In TMS (Figure 26.9), carbon is bonded to a *less* electronegative atom (silicon), which is why the electron density around carbon is higher in TMS than in almost all other organic compounds.

We describe the proton as **deshielded** if the electron density is reduced because of the presence of an electronegative atom such as oxygen; this produces a larger chemical shift.

• Figures 26.11, 26.12, and 26.13 show examples of ¹³C NMR spectra.

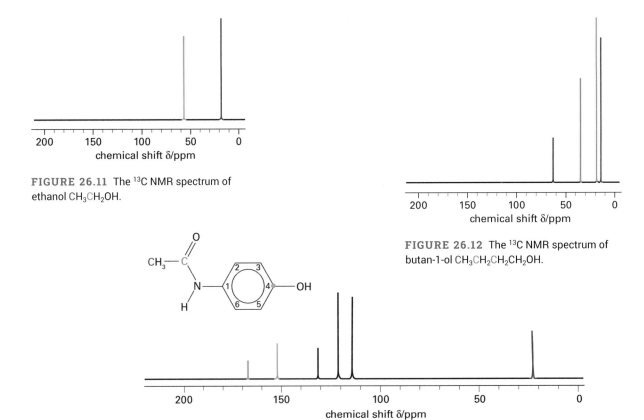

FIGURE 26.11 The ¹³C NMR spectrum of ethanol CH_3CH_2OH.

FIGURE 26.12 The ¹³C NMR spectrum of butan-1-ol $CH_3CH_2CH_2CH_2OH$.

FIGURE 26.13 The ¹³C NMR spectrum of paracetamol. Owing to the symmetry of the positions round the benzene ring, carbon atoms 2 and 6 are identical, as are carbon atoms 3 and 5, hence there are only three peaks in the region of 100 to 135 ppm. Note that the amide group ($NHCOCH_3$) is able to rotate about the bond between the nitrogen atom and the benzene ring.

26.5 ¹H NMR (PROTON NMR)

- The **number of peaks** indicates the number of different types of proton: **chemically equivalent** protons contribute to the same peak.

- The **chemical shift δ** (measured in parts per million, ppm, relative to TMS) depends on the detailed environment of the particular proton, see Table 26.3.

- We can find two other pieces of information from ¹H NMR spectra which are not available from ¹³C NMR.

  - First, the area under each peak (the **peak area**) is **directly proportional to the number of equivalent protons** contributing to the peak. We can find the peak area by numerical integration of the peak, which is shown as an **integration trace**. Hence we can find the *relative* numbers of equivalent protons contributing to each peak. Figure 26.14 shows the ¹H NMR spectrum for ethanol.

    Note that the peak area is **not** directly related to the number of equivalent carbons in ¹³C NMR, nor is splitting normally observed in ¹³C NMR because of the experimental technique used (broadband proton decoupling).

  - Second, there is a detailed **splitting** of each peak, described in Section 26.6.

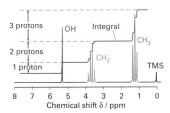

FIGURE 26.14 The ¹H NMR spectrum of ethanol. (The OH peak shows no splitting because of rapid proton exchange.)

TABLE 26.3 Approximate chemical shifts and corresponding proton environments. Here R is an alkyl group.

Type of proton	Chemical shift δ/ppm
RCH_3	0.8-1.2
R_2CH_2	1.2-1.4
R_3CH	1.5-1.7
(R–C(=O)–CH₃)	2.0-2.3
(benzene–CH₃)	2.3-2.5
$R_2C=CH_2$	4.5-5.0
$R_2C=CHR$	5.0-5.5
(benzene–H)	6.8-7.8
(R–C(=O)–H)	9.7-10.0

A general, albeit rough, rule of thumb is that the ¹³C NMR chemical shift is twenty times that of the ¹H NMR chemical shift (the **20x rule**).

As was the case for ¹³C chemical shifts, you should consider this table to be a useful guide rather than a definitive list.

26

26.6 SPIN–SPIN SPLITTING (SPIN–SPIN COUPLING) IN ^{1}H NMR SPECTRA

The splitting discussed pre-university applies to protons that are exactly **three bonds apart**. At university, you will be given more detailed analysis of the splitting.

This is strictly only true when the difference in chemical shifts is significantly larger than the spin–spin splitting.

- The presence of a proton (or protons) on **adjacent, non-equivalent** carbon atoms will create a very small magnetic field, which might add to or subtract from the external field, causing a splitting in the peak.
- The splitting depends on the *number* of adjacent protons, *n*. The peak will be split into (*n*+1) closely spaced peaks, which we call the **(n+1) rule**:
  - If there are **zero** adjacent protons the peak is a **singlet**.
  - If there is **one** adjacent proton the peak is a **doublet**.
  - If there are **two** adjacent protons the peak is a **triplet**.
  - If there are **three** adjacent protons the peak is a **quartet**.
- We can find the relative intensities using **Pascal's triangle**, as shown in Figure 26.15.
  - A doublet will be a **1:1 doublet**.
  - A triplet will be a **1:2:1 triplet**.
  - A quartet will be a **1:3:3:1 quartet**.

Figure 26.14 shows the ^{1}H NMR spectrum of ethanol. The CH_3 peak is a triplet because of the two adjacent protons, and the CH_2 peak is a quartet because of the three adjacent protons. Such a pattern is typical when an ethyl (CH_3CH_2) group is present.

Figure 26.16 shows the ^{1}H NMR spectrum of 1,1,2-trichloroethane. The CH_2 peak is a doublet because of the one adjacent proton, and the CH peak is a triplet because of the two adjacent protons.

See also Figure 22.5 for the splitting seen in ethanal CH_3CHO.

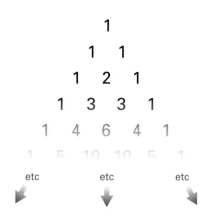

FIGURE 26.15 Pascal's triangle.

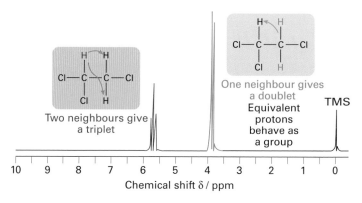

FIGURE 26.16 The ^{1}H NMR spectrum of 1,1,2-trichloroethane.

Periodic Table

Period	Group 1	2	3	4	5	6	7	8	9	10	11	12	13	14	15	16	17	18
1	1 H Hydrogen 1.0																	2 He Helium 4.0
2	3 Li Lithium 6.9	4 Be Beryllium 9.0											5 B Boron 10.8	6 C Carbon 12.0	7 N Nitrogen 14.0	8 O Oxygen 16.0	9 F Fluorine 19.0	10 Ne Neon 20.2
3	11 Na Sodium 23.0	12 Mg Magnesium 24.3											13 Al Aluminium 27.0	14 Si Silicon 28.1	15 P Phosphorus 31.0	16 S Sulfur 32.1	17 Cl Chlorine 35.5	18 Ar Argon 39.9
4	19 K Potassium 39.1	20 Ca Calcium 40.1	21 Sc Scandium 45.0	22 Ti Titanium 47.9	23 V Vanadium 50.9	24 Cr Chromium 52.0	25 Mn Manganese 54.9	26 Fe Iron 55.8	27 Co Cobalt 58.9	28 Ni Nickel 58.7	29 Cu Copper 63.5	30 Zn Zinc 65.4	31 Ga Gallium 69.7	32 Ge Germanium 72.6	33 As Arsenic 74.9	34 Se Selenium 79.0	35 Br Bromine 79.9	36 Kr Krypton 83.8
5	37 Rb Rubidium 85.5	38 Sr Strontium 87.6	39 Y Yttrium 88.9	40 Zr Zirconium 91.2	41 Nb Niobium 92.9	42 Mo Molybdenum 96.0	43 Tc Technetium (98)	44 Ru Ruthenium 101.1	45 Rh Rhodium 102.9	46 Pd Palladium 106.4	47 Ag Silver 107.9	48 Cd Cadmium 112.4	49 In Indium 114.8	50 Sn Tin 118.7	51 Sb Antimony 121.8	52 Te Tellurium 127.6	53 I Iodine 126.9	54 Xe Xenon 131.3
6	55 Cs Caesium 132.9	56 Ba Barium 137.3	57 La Lanthanum 138.9	72 Hf Hafnium 178.5	73 Ta Tantalum 180.9	74 W Tungsten 183.8	75 Re Rhenium 186.2	76 Os Osmium 190.2	77 Ir Iridium 192.2	78 Pt Platinum 195.1	79 Au Gold 197.0	80 Hg Mercury 200.6	81 Tl Thallium 204.4	82 Pb Lead 207.2	83 Bi Bismuth 209.0	84 Po Polonium (209)	85 At Astatine (210)	86 Rn Radon (222)
7	87 Fr Francium (223)	88 Ra Radium (226)	89 Ac Actinium (227)	104 Rf Rutherfordium (267)	105 Db Dubnium (270)	106 Sg Seaborgium (269)	107 Bh Bohrium (270)	108 Hs Hassium (270)	109 Mt Meitnerium (278)	110 Ds Darmstadtium (281)	111 Rg Roentgenium (281)	112 Cn Copernicium (285)	113 Nh Nihonium (286)	114 Fl Flerovium (289)	115 Mc Moscovium (289)	116 Lv Livermorium (293)	117 Ts Tennessine (293)	118 Og Oganesson (294)

58 Ce Cerium 140.1	59 Pr Praseodymium 140.9	60 Nd Neodymium 144.2	61 Pm Promethium (145)	62 Sm Samarium 150.4	63 Eu Europium 152.0	64 Gd Gadolinium 157.3	65 Tb Terbium 158.9	66 Dy Dysprosium 162.5	67 Ho Holmium 164.9	68 Er Erbium 167.3	69 Tm Thulium 168.9	70 Yb Ytterbium 173.0	71 Lu Lutetium 175.0
90 Th Thorium 232.0	91 Pa Protactinium 231.0	92 U Uranium 238.0	93 Np Neptunium (237)	94 Pu Plutonium (244)	95 Am Americium (243)	96 Cm Curium (247)	97 Bk Berkelium (247)	98 Cf Californium (251)	99 Es Einsteinium (252)	100 Fm Fermium (257)	101 Md Mendelevium (258)	102 No Nobelium (259)	103 Lr Lawrencium (262)

INDEX